Dr. Simoudis paints a full and nuanced picture of the  connectivity, and electrification. A must read for any tant disruption or who will be impacted by it.

DR. STEVEN E. KOONIN
FORMER UNDER SECRETARY FOR SCIENCE, US DEPARTMENT OF ENERGY,
DIRECTOR, NYU CENTER FOR URBAN SCIENCE AND PROGRESS

In this well-researched and easily readable book, Dr. Simoudis makes a strong case for the importance of exploiting big data as autonomous vehicles are employed in next-generation mobility solutions. The book presents an in-depth analysis of the opportunities big data offers to the automotive industry and provides a compelling set of recommendations

DR. ULRICH QUAY
MANAGING DIRECTOR, BMW IVENTURES

Dr. Simoudis presents a crystal clear, concise and well researched perspective of the data revolution facing the transportation industry. We are in the midst of a mobility renaissance where data from users and smart sensors will create new opportunities with lower barriers to entry, thus driving disruption in the automotive space. This book gives you technical, business and venture investment insight, and provides recommendations and perspective on best practices to consider in creating an innovation culture!

JOHN ABSMEIER
INNOVATION LEADER, AUTONOMOUS DRIVING EXPERT AND
VICE PRESIDENT OF SMART MACHINES AT SAMSUNG ELECTRONICS

Thoughtful insight and advice for leveraging the disruption in the automotive world — but just as applicable to other areas that are ripe for disruption.

DR. WILLIAM MARK
PRESIDENT, INFORMATION AND COMPUTING SCIENCES, SRI INTERNATIONAL

The next vanguard in the automobile business is likely a company that understands both hardware and software. The only way to live effectively between those two polar forces is to capture and manage data effectively. Dr. Simoudis understands how companies can build their operations around this data mindset and this book contains many of the recipes.

REILLY BRENNAN
EXECUTIVE DIRECTOR, REVS PROGRAM, STANFORD UNIVERSITY

Autonomous Connected Electrified vehicles have the potential to spark the greatest positive change in our society since the industrial revolution. The Big Data Opportunity In Our Driverless Future provides an important perspective on the role of big data and machine intelligence technologies in such vehicles and the infrastructure they operate on, the transportation experiences and opportunities such vehicles will enable for consumers and industries, and the societal challenges they will mitigate.

ROBBIE DIAMOND
CEO AND FOUNDER, SECURING AMERICA'S FUTURE ENERGY (SAFE)

Dr. Simoudis' deep and thoughtful analysis of the automotive industry and its ecosystem, the rise of big data and analytics, and the tectonic change that it represents to this storied industry makes his book a pleasure to read and will likely become mandatory material for the global executives, business students, and entrepreneurs alike. The Big Data Opportunity In Our Driverless Future clearly describes the massive challenges the industry faces and provides actionable recommendations that any company hoping to transform itself can adopt. Anyone interested in understanding the profound societal transformations that can arise when insights and decisions are based on data will surely enjoy this book.

ASHOK SRIVASTAVA
CHIEF DATA SCIENTIST, VERIZON CORPORATION

Dr. Simoudis delivers a powerful message: Tomorrow's customer will want transportation not just in the form of a car in the driveway but also as an on-demand service. This megashift will trigger enormous changes in society. To win in the future, automotive companies will need to master three new weapons: Big data, machine intelligence and an innovation culture that feels just like a start-up. Simoudis' insightful and concise book, The Big Data Opportunity, shows us how.

MICHAEL DUNNE
AUTHOR: AMERICAN WHEELS, CHINESE ROADS
PRESIDENT: DUNNE AUTOMOTIVE LTD.

The Big Data Opportunity In Our Driverless Future delivers insights and intelligence with an exceptional level of precision and focus, traversing the strategic to tactical implications of big data on the future of autonomous vehicles. The book reaches its ambitious goal of educating and guiding readers into the key issues surrounding driverless or autonomous vehicles, highlighting the role of big data and analytics in the rapid evolution of the automotive industry and transportation today

LOUIS COLUMBUS
DIRECTOR, GLOBAL CLOUD PLATFORM MANAGEMENT
AND CLOUD SERVICES, FORBES CONTRIBUTOR

THE
BIG DATA
OPPORTUNITY
IN OUR
DRIVERLESS
FUTURE

The Big Data Opportunity in Our Driverless Future

Copyright © 2017. All Rights Reserved.

No part of this publication may be reproduced, stored in a retrieval system or transmitted, in any form or by any means—electronic, mechanical, photocopying, recording or otherwise—without prior written permission from the publisher, except for the inclusion of brief quotations in a review.

For information about this title or to order other books and/or electronic media, contact the publisher:

Corporate Innovators, LLC

www.corporateinnovation.co

info@corporateinnovation.co

Printed in the United States of America

Cover and Interior design: 1106 Design

Publisher's Cataloging-In-Publication Data
(Prepared by The Donohue Group, Inc.)

Names: Simoudis, Evangelos.

Title: The big data opportunity in our driverless future / Evangelos Simoudis, Ph.D.

Description: Menlo Park, CA : Corporate Innovators, [2017] |
Includes bibliographical references and index.

Identifiers: LCCN 2016963485 | ISBN 978-0-9980677-1-1 | ISBN 978-0-9980677-0-4 (ebook)

Subjects: LCSH: Autonomous vehicles—Technological innovations. | Autonomous vehicles—Data processing. | Big data—Industrial applications. | Intelligent transportation systems. | Automobile industry and trade—Technological innovations.

Classification: LCC TL152.8 .S56 2017 (print) | LCC TL152.8 (ebook) | DDC 629.222—dc23

CONTENTS

PREFACE

Transportation. Communication. Entertainment. The common threads that run through all of these familiar industries are disruption and change.

The modern transportation industry—including vehicle manufacturers, parts suppliers, distributors, retailers, service, and infrastructure providers—is in the early stages of a sweeping transformation, drawing obvious parallels with the technological, social, and economic upheaval that has overturned and redrawn the communication and entertainment industries. For the automotive establishment, these changes likely will result in dramatically different business models for those companies that hope to survive and prosper as next-generation purveyors of mobility and related services.

Evangelos Simoudis, a veteran Silicon Valley venture capitalist and one-time entrepreneur, offers a thoughtful prescription for auto industry incumbents: Embrace Big Data as the core of what Simoudis describes as a Startup-Driven Corporate Innovation Strategy.

Drawing from his own experience with technology startups and big data, Simoudis lays out a blueprint for survival and success, one that envisions a constantly evolving business cycle that effectively manages risk, disruption, innovation, and, ultimately, change.

There is a practical bent to his vision and recommendations. The theory presented here is harnessed specifically to the rapid evolution of the industry, and of mobility in general, from an individual ownership model to an access model—in other words, on-demand transportation as a service. Hand in hand with this movement is an accelerating shift to what Simoudis labels ACE vehicles—autonomous, connected, electrified. How these pieces fit together in the industry's broader strategy and how incumbents can adapt the most appropriate and relevant lessons from successful startups to help drive organizational change and innovation are at the heart of Simoudis's message.

In order to innovate continuously, he argues, corporations not only have to learn how to work more effectively *with* startups, they also have to work *like* startups. That means they need to incorporate innovative thinking in their core business, in emerging opportunities that are either extensions of existing business or new ventures, and finally in potentially disruptive ventures. It's not enough to have a well-funded R&D operation at home or a Silicon Valley "innovation outpost," as many automakers and suppliers have established over the past 10 to 15 years. Incumbents have to do a better job of knitting together what are often disparate strategic pieces while working proactively to build and promote a corporate culture in which innovation is embedded as a core value.

More specifically, incumbents need to make big data, coupled with machine intelligence, a cornerstone of their strategic thinking. Simoudis urges companies to foster a data-sharing culture married to a collaborative ecosystem that connects them not only with established suppliers but also with tech startups that bring a fresh perspective and new solutions to emerging challenges such as autonomous driving. Those partnerships include ride services companies such as Uber and Lyft that are fast becoming pivotal players in the sharing/on-demand economy.

Underlying all is big data, which Simoudis describes as "a key ingredient in the future of mobility," from autonomy to more effective personalization and monetization of mobile services such as e-commerce and infotainment. His message is clear: Technology itself can be a useful tool and enabler, but it's not enough. Companies must learn to exploit big data: To collect and manage it, to fuse and analyze it, to model and communicate it. Finally, Simoudis urges the creation and industry-wide adoption of a unified data architecture—a standard operating system that is open, adaptable, and extensible, one that combines big data and computing power and can be shared across the partner ecosystem.

These are valuable and insightful solutions, intended to help executives map out winning strategies to better manage the Darwinian risk/disruption/innovation/change cycle that has characterized the auto industry for more than a century. If effectively developed and implemented, the exploitation of big data, along with the adoption of a startup-driven corporate innovation strategy, will provide incumbents with critical survival skills and tools that might have altered the outcomes for other now-defunct or diminished companies in other rapidly evolving industries.—*Paul Lienert*

1.0
INTRODUCTION

E arly in 2016, GM reportedly paid close to $1 Billion to acquire Cruise Automation, a 40-employee, Silicon Valley-based, venture-funded startup that was developing driverless vehicle technology. A few months earlier Toyota had announced that it would invest $1 billion on artificial intelligence research, one of the key technologies making driverless vehicles possible. From Detroit to Germany, Japan, and Korea, there is an amplifying conversation about the magnitude, extent, and timing of the automotive industry's disruption that will be the result of the introduction of autonomous and driverless vehicles. This disruption will come as we try to address important societal and urban challenges and changes. It will be the result of changing attitudes toward car ownership, technology innovations, and business model innovations. The **electrified, autonomous**, and ultimately *driverless* **connected** vehicle will be offered in conjunction with a variety of on-demand Mobility Services under a hybrid model that blends car ownership with on-demand car access. Big data coming from inside and outside the autonomous vehicle and machine intelligence technologies used for the exploitation of this data are key ingredients in these next-generation vehicles. They also offer a unique, and currently overlooked, opportunity as we proceed toward a driverless future. To survive and ultimately thrive in a future where new mobility models become the norm, incumbent automakers must adopt different approaches to innovation. They must invest heavily and over a long period in technologies in which they have little or no competence, such as big data and machine intelligence, and combine technology innovations with the right business model innovations like startups in Silicon Valley, Israel, and China are doing routinely.

What is causing a company like GM to pay such a sum to acquire a tiny startup, for Toyota to commit such a large amount on artificial intelligence research, and for almost every major automaker to establish a technology center in Silicon Valley with frequent visits by each company's CEO? Try to imagine riding in a driverless vehicle during the rush-hour in a city like Los Angeles or Beijing. In addition to being freed from the stress caused by rush-hour traffic, and using the commuting time in any way you like, you will not have to worry about the issues associated with arriving at your destination: Where can I find parking? How close to my destination will I be if I park in a particular location? Will I need to drive to arrive at my next meeting on time? Answers to these questions impact the customer experience and will impact the business of the incumbent automotive and transportation industries. With these actions, the automotive industry incumbents are attempting to acquire or develop the technology that is rapidly becoming table stakes for next-generation vehicles and thus address an emerging disruption risk. They are also trying to respond to the new technology challenges and the customer experience innovations that a driverless future will bring. But billion-dollar acquisitions and investments, though impressive, are not sufficient by themselves to address the disruption risk the incumbent automotive industry is facing. The industry will need to:

- *Transform* from designing and manufacturing vehicles, to being in the insights business and offering transportation solutions that will be consistent with recent consumer model changes;
- *Adopt* a new corporate innovation strategy and make the appropriate organizational and cultural changes in order to make the transformation successful;
- *Designate* big data and its exploitation using machine intelligence as a strategic imperative around electrified autonomous vehicles and on-demand mobility services. Several automakers in addition to Toyota are starting to recognize that machine intelligence is necessary for analyzing the sensor data generated by driverless vehicles and using the results for the navigation of such vehicles. However, they miss that machine intelligence is equally important for the exploitation of many other types of big data that is generated in and around next-generation vehicles and by the plethora of mobility services. The industry must rapidly *develop* expertise and critical mass in these areas because they represent a strategic opportunity where incumbent automakers have the potential to create a competitive advantage and lead in a driverless future.

In this book, we will:

- Explore some of the societal and urban challenges that are driving innovation in the automotive industry;
- Discuss the technology and business model innovations being developed to address these challenges with a focus on big data and its exploitation using machine intelligence;

- Describe the critical role of big data and machine intelligence in next-generation mobility, and provide a framework for organizing the relevant big data so that it can be effectively exploited using machine intelligence to create value;
- Identify the companies that have initiated the disruption through their innovations and their visions on personal transportation and analyze their characteristics;
- Present recommendations for how the automotive industry's incumbents can take advantage of this dynamic new landscape, by re-thinking their approach to innovation, particularly over-the-horizon innovation, and designating big data a strategic imperative.

2.0
A FEW FACTS ABOUT THE AUTOMOTIVE INDUSTRY

B efore discussing the factors and innovations that are leading to the automotive industry's disruption, it is useful to present a few facts about the industry itself.

The automotive industry (approximately $1T in annual sales) is dominated by a group of 14 very large automotive Original Equipment Manufacturers (OEMs), with their several dozen brands, shown in Figure 1.

The reason[1] there are so few automotive OEMs is because starting one today is extremely capital-intensive, creating high barriers to entry. For example, see the capital raised by the now defunct Fisker Automotive, the capital purported to have been raised by Faraday Future[2] and NextEV, and the capital raised by Tesla Motors (to date, and including the company's IPO). Suffice it to say that Tesla Motors was the first automotive OEM to go public since the Ford Motor Company went public in 1956.

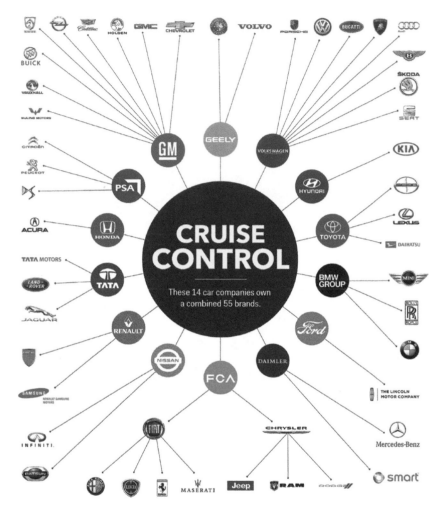

FIGURE 1: THE LARGEST AUTOMOTIVE OEMs AND THEIR BRANDS. FIGURE COURTESY OF **SKYE GOULD/BUSINESS INSIDER**.[3]

New car sales worldwide in 2015 were 83 million vehicles and represent $500B of the market. Of these new cars, about 17 million were sold in the US. In addition, during 2015 in the US alone, about 38 million used cars were sold. Despite its capital intensity, automotive manufacturing is a low-margin business.[4] Of the automotive OEMs Toyota is best in class in terms of EBITDA margins.

Over the years OEMs have transitioned from being vertically integrated companies to highly efficient integrators of components in car platforms they define and own. Today OEMs are responsible for the design, manufacturing and distribution of the vehicle. Their supply chains are global and are optimized through constant incremental improvements, giving them cost advantages a new entrant cannot achieve. Making radical changes to this supply chain is difficult, if not impossible, because in an effort to improve their margins, OEMs enter into long-term agreements with their suppliers. Tesla, for example, has recently started finding out what it takes to create a global supply chain and manufacturing process that can provide the worldwide market with hundreds of thousands of vehicles.

While initially these were hardware-only platforms, today's cars can be thought of as consisting of a software platform, of mostly embedded and proprietary software that controls major functions of the car, and a hardware platform. According to a report published by the Center for Automotive Research, the automotive industry spends $100B/year on R&D, which equates to $1,200 per vehicle produced.

As is shown in Figure 2, most of this investment is made by the OEMs to improve the car's platform and the elements that control this platform's performance, make it safer, more efficient, etc. and on manufacturing and distribution process improvements. The suppliers (Tier 1, Tier 2, and so on) are responsible for the R&D around the components they provide to the OEM for each platform. For example, Bosch, a Tier 1 automotive supplier of infotainment systems to companies like BMW

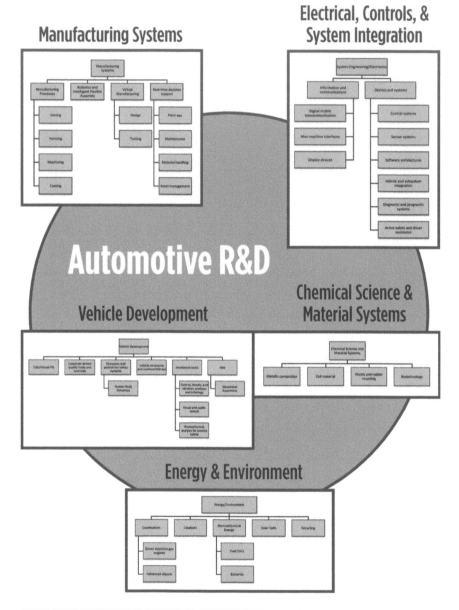

FIGURE 2: TYPICAL AUTOMAKER'S R&D AREAS OF FOCUS. FIGURE COURTESY OF CENTER FOR AUTOMOTIVE RESEARCH.

and Daimler, is responsible for conducting the R&D around the systems it supplies to its partner OEMs. But this R&D is still based on the principles of a gas-powered engine and associated powertrain with all of their complexities. This type of technology-centered R&D investment is characteristic of companies aiming at innovations that enable them to sustain and improve on their existing business models, i.e., improve their profit margins while extending the viability of the business model, processes, and supply chains, in this case, models, processes, and supply chains related to gas-powered vehicles.

But as my good friend Lou Kerner,[5] a Partner at Flight VC, helped me realize—when we talk about the automotive value chain, we typically think about OEMs, their suppliers and dealers, and yet—the automotive value chain should actually be organized into two broad parts: 1) vehicle manufacturing, distribution, and sales; and 2) vehicle use.

Vehicle manufacturing, distribution, and sales (shown in Figure 3) includes the OEMs, the very large number of hierarchically organized suppliers, the logistics companies that are responsible for moving the parts and bringing the new vehicles closer to the consumer (distribution), and the thousands of dealers that are selling and servicing the vehicles (today there are about 17,000 new-car dealers by the most recent industry estimates).[6]

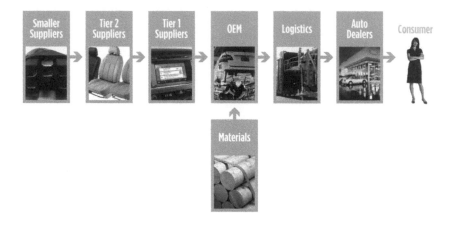

FIGURE 3: THE VEHICLE MANUFACTURING, DISTRIBUTION, AND SALES PART OF THE AUTOMOTIVE VALUE CHAIN. FIGURE COURTESY OF EVANGELOS SIMOUDIS.

The Vehicle Use Value Chain is shown in Figure 4. This value chain is:

- **Fragmented**, with each area having its own ecosystem of established companies and increasingly more startups. Some areas, even if new, have low barriers to entry (e.g., driving analytics), while others have high barriers because they are more capital intensive (e.g., ridesharing) or require regulatory approvals (e.g., vehicle insurance).

- **In flux**, with new categories, entrants, and disruptors, particularly in the areas of Driving Services and Mobility Services. For example, infotainment, expected to be the next battleground of the automotive value chain, used to be dominated by established hardware suppliers that have been part of the car manufacturing, distribution, and sale value chain. As cars are becoming increasingly connected, companies such as Google and Apple have entered this arena through the vehicle use value chain based on their software and content solutions. Startups like Truvolo and Mojio, and many others, are also entering through a combination of hardware/software solutions. Finally, automotive OEMs are also getting into this space through acquisitions of startups.[7]

- **Diverse**, with more participating industries, e.g., insurance, utilities, energy, telco. For example, the introduction of electrified vehicles automatically made the utility industry part of this value chain; the introduction of broadband connectivity made the telco industry part of this value chain; and the use of personal navigation made companies like Google, Apple, and HERE parts of the value chain.

- **Dominated by startups**, particularly the areas of Driving Services and Mobility Services, where new technologies, and particularly software, big data, semiconductor, and communication technologies become the key ingredients for defining these services and provide opportunities for disruption. For example, a database that I maintain of startups working in the vehicle use value chain thus far contains more than 500 entries. We need to also note that startups offering ridesharing, carsharing, and goods-delivery services have attracted the lion's share of attention and venture capital funding (almost 2/3 of the total funding that has gone into startups in the vehicle use value chain). As we will discuss in a later chapter, many of these startups are consumers and/or producers of big data, and several are innovating by exploiting it using machine intelligence.

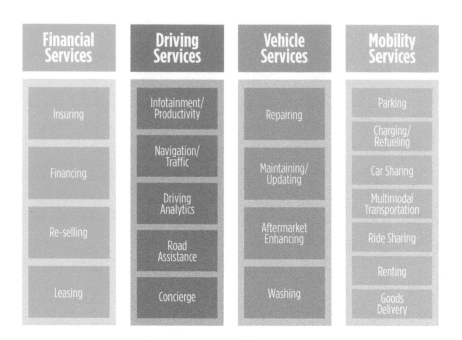

FIGURE 4: THE VEHICLE USE PART OF THE AUTOMOTIVE VALUE CHAIN. FIGURE COURTESY OF EVANGELOS SIMOUDIS.

Automotive OEMs had previously entered parts of the vehicle use value chain, particularly with their financial services units. More recently, a few OEMs have started entering other parts of this value chain, as they start to recognize the potential size of markets such as ridesharing. For example, Tesla has established charging stations, and BMW has established the DriveNow carsharing division.

3.0
DRIVERS OF AUTOMOTIVE DISRUPTION

nnovative transportation solutions to societal and urban challenges, primarily developed by startups that are exploiting technology and business model innovations, are leading to a big shift away from car ownership and have the potential of disrupting the incumbent automotive industry, even though in the short term automakers may continue to report record sales.

3.1 SOCIETAL AND URBAN CHALLENGES

According to a UN report in 2014,[8] more people lived in urban areas (54%) than in rural areas. For example, many young people these days want to live in cities. An important consequence of this increasing urbanization is the move by several large corporations,[9] such as Boeing, GE, McDonald's, and others, from their suburban headquarters, where they had been located for decades, to the center of cities in order to be in the position to hire younger employees. By 2050, an additional 2.5B people are projected to live in urban areas. Today's megacities (metropolitan areas with a population of more than 10M people)—New York, Mumbai, Beijing, Mexico City, São Paulo, and several others—will continue to grow, and many new ones will be added, mostly in Asia and Africa. As research centers such as the Center for Urban Science and Planning (CUSP)[10] and the Future Cities Laboratory are demonstrating, megacities must address several different problems. But three are directly related to transportation, particularly personal car use, and one where transportation is a significant contributor.

- **Problem 1**: Productivity Loss. We are starting to realize that as each megacity's population continues to grow, particularly in megacities with low population density like Los Angeles or Bangkok, its road and parking infrastructures cannot expand proportionately in a way that will support the prevailing car ownership and use models. We spend too much time driving to commute to work or home, and when we arrive at our destinations we have the wrong frame of mind to be productive. By looking at the commute times in megacities like Mexico City and Beijing (shown in Figure 5) and the productivity loss that is associated with long commutes, we should understand that this is already a serious problem. In fact, as cities like Singapore are working to reduce the amount of space they allocate to transportation infrastructure and shift it to residential and recreational uses in order to further increase their population density, the car ownership-centric model will become even more unfeasible. The available data implies that this problem is more acute in Asia and Africa, rather than North America or Europe.
- **Problem 2**: Pollution. Even more important to loss in worker productivity is the negative environmental impact[12] of gas-powered vehicles in urban (and suburban) settings, as this is exhibited through the increasingly poor air quality in megacities[13] due to carbon emissions. Consider the air quality in megacities like Beijing, Jakarta, and Mexico City.

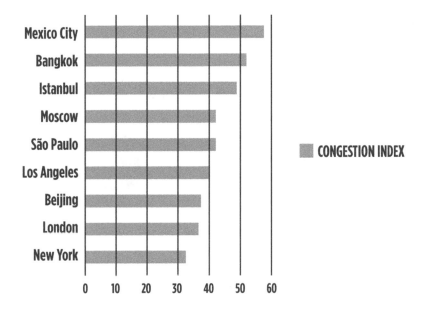

FIGURE 5: MEGACITIES CONGESTION INDEX. FIGURE COURTESY OF EVANGELOS SIMOUDIS USING DATA PROVIDED BY TOMTOM.[11]

- **Problem 3**: Climate change. Many have argued far more eloquently about the impact of gas-powered transportation on greenhouse gas emissions that spur climate change.[14] Though now a little dated (they include data up to only 2006), figures published by the US Department of Transportation show that in the US, transportation contributes 28% of greenhouse gases, with more than 50% of that coming from passenger cars and light-duty vehicles, a category that includes sport utility vehicles, pickup trucks, and minivans. If we are to look at available data from Asia, we conclude that transportation has been responsible for 19% of emissions and is projected to be 31% by 2030.

- **Problem 4**: Employment, cost of living, and salaries. During this
period of continued economic expansion, many consumers, and
particularly Millennials, are still struggling to find permanent jobs.
Several years after the 2009 recession, there is an employment gap.[15]
This is particularly the case for entry-level jobs, which typically do
not provide Millennials with income that is adequate and appropriate for the cost of living in the locations where these jobs exist, e.g.,
San Francisco and NY. More broadly, even for non-entry-level jobs,
wages have not grown in terms of real purchasing power. At the
same time, Millennials carry higher debt due to their student loans.
Income disparity is fueling the **Sharing Economy**[16] (also known
as the **Collaborative Economy**)[17] that is broadly used particularly
by Millennials to benefit from its services but also to contribute
income-producing services. A recent study[18] found that as a result of
these trends, fewer young people are getting driver's licenses.

Unfortunately there is not a single "silver bullet" solution to these problems. We can't solve the productivity-loss problem by just building more
roads and parking garages. We can't fix the pollution problem by just low-
ering vehicle CO_2 emissions. We can't address the climate change problem
by just discouraging personal cars. And while the continuously improving
economic environment is leading to the creation of more jobs, leading
Millennials to purchase cars,[19] the reality is that many, particularly those
living in cities, are starting to forgo car ownership.

Addressing these issues will require:

1. Application of innovative technologies, combined with innovative
   business models.

2. A major shift in consumer behavior relating to car ownership and use.

These solutions are also drivers of disruption in the automotive industry as well as the energy industry.

3.2 TECHNOLOGY INNOVATIONS: AUTONOMOUS CONNECTED ELECTRIFIED VEHICLES

Many different technology innovations are starting to reshape the automotive industry, but none can be as transformative as the **Autonomous Connected Electrified (ACE) vehicle.** Of the problems presented in the previous section, over time vehicle electrification has the potential to address the pollution and climate change problems. Over the next 30–40 years vehicle autonomy combined with on-demand Mobility Services could address the productivity and pollution problems, as well as other issues, such as reducing accidents, improve the utilization of the existing transportation infrastructure, and providing mobility options for the elderly and handicapped.

While acknowledging the role of vehicle electrification in having the potential to address the pollution and climate change problems, this is a topic outside the book's scope. I will only note that even though estimates vary, I expect more than 25% of the vehicles to be electric, or electrified, by 2025–30 (up from 5% today) and be able to meet or exceed the emission standards targeted for 2025. Most incumbent OEMs have either already introduced or have announced the future introduction of several electrified passenger car models in their lineup that are capable of driving increasingly longer distances on a single charge. In addition we are starting to see the introduction of other types of electrified vehicles, such as city buses and short-haul trucks. We are expecting that the majority of the future autonomous vehicles will be electrified.

We should note that in order to effectively address the societal and urban challenges stated earlier, the automotive industry must think not only about how electric vehicles are being used but also how other industries are being affected from their use and, even more importantly, about the environmental impact of their manufacturing and disposal of spent battery cells. For example, a report by the National Academy of Sciences[20] describes how the emissions created during the manufacturing of an electric vehicle and its fuel are far greater than the emissions of operating an average gasoline-powered car.

A short historical perspective of autonomous vehicles

While we now see articles about self-driving or autonomous vehicles on a daily basis, automakers and universities have been experimenting with such vehicles for many years. In fact, limited experimentation started as early as the 1950s, and the first experimental autonomous cars were road-tested in the US and Europe in the 1980s. As in many other fields, the lack of sufficient computing power, broadband connectivity, the availability of sensors, particularly sensors at cost that can be afforded and in sizes that can be accommodated onboard the vehicle, confined these efforts to research laboratories. However, many lessons were learned.

In 1989 Dean Pomerleau, then a student at Carnegie Mellon University (CMU), wrote a seminal paper,[21] as part of his Ph.D. research, which described how a neural network could be trained and used for vehicle navigation. It was the first demonstration of how big data, which in that case was image data and range data, could be used by autonomous vehicles.

In 2004 DARPA announced the Autonomous Vehicle Grand Challenge.[22] Over the following three years DARPA held three competitions in which teams were able to demonstrate increasingly sophisticated autonomous vehicles that were able to travel longer distances in a variety of driving conditions, including challenging urban settings. By then many of the technology problems that were limiting earlier efforts had started to be solved. Two years after the last DARPA competition, in 2009, Google started its autonomous car effort. Sebastian Thrun,[23] who had led Stanford's teams in DARPA's competitions, headed it.

DARPA's Grand Challenge competitions made many contributions to the autonomous vehicle efforts. They succeeded in focusing the attention of many research teams of that era on the problem of autonomous driving. They demonstrated:

1. That a certain level of autonomous vehicle performance was feasible with the technology of the time. It was after these competitions were completed that companies like Google and some startups began their efforts on autonomous vehicles.

2. The importance of using a multitude of sensors and in general the relation between a vehicle's technological sophistication and its ability to negotiate a particular course. Though the vehicles that participated in the competitions had to be much larger than today's autonomous cars to be able to carry all the sensors and computing resources they needed to accomplish their task, by 2004 sensors had become small enough and portable computing powerful enough to make such competitions viable.

3. The importance of mapping in autonomous driving. The first to demonstrate the significance of mapping was the team from CMU.

The participating teams also demonstrated that neural networks can be used effectively in navigational systems. As we will see in the next chapter, this led to the eventual use of deep learning networks.

4. The direct correlation between computing power on-board the vehicle and driving speed on the tested courses.

5. The importance of having an always-on communication connection to the vehicle. While the vehicles that participated in the competitions had a significant amount of computing power compared to prior efforts in autonomous driving, it was still necessary to perform certain computations off-board and communicate the results of these computations to the vehicle.

6. The difficult problems that must be solved when autonomous vehicles operate in open-ended environments vs. predefined closed courses.

Levels of driving automation

There exists significant confusion about the terms "self-driving," "autonomous," and "driverless," which are used interchangeably by the press. Automakers typically organize driving automation into **six levels**. **Level 0** refers to *no automation*; the human driver has control of every aspect of the vehicle at all times. At the other end, **Level 5** refers to *complete automation,* where every aspect of the vehicle is controlled by the driving system (software, hardware, data)—the so-called *driverless vehicle*, such as the experimental Google Cars.[24]

The other four levels of driving automation consist of:

- **Level 1**: Vehicles include driver-assistance functionality that is used for steering, braking, and acceleration using data about the driving environment. The driver is still responsible for monitoring the driving environment and controlling most vehicle features.

- **Level 2**: In addition to braking and acceleration, the driving system becomes responsible for lane centering, implying that the driver can take his hands off the steering wheel and his foot off the pedals.
- **Level 3**: At this level all of the vehicle's safety functions are taken over by the driving system. At this level the vehicle can be characterized as *self-driving*. However, it is expected that the human driver will always be able to intervene and take over when necessary.
- **Level 4**: Vehicles at this level are autonomous. They can take over vehicle functions and monitor road conditions and overall environment conditions for an entire trip, even if the human driver does not respond to a request to intervene and take over when asked.

Today a growing number of OEMs are already selling cars with Level 1, Level 2, and Level 3 driving automation[25] (such as Tesla's Autopilot feature). More vehicles with Level 3 and 4 driving automation will be introduced in the next 3–5 years.[26] As of now, 33 corporations[27] are working on autonomous vehicles, including many major automakers, many newer automakers such as Tata Elxsi and several of their Tier 1 suppliers. In addition to Google, many companies outside the established automotive value chain such as Baidu are also working on autonomous vehicle technology. Automotive incumbents like Mercedes and BMW are moving progressively from Level 2 to Level 3 driving automation and then on to Level 4 and Level 5, whereas Google is moving directly to vehicles with Level 4 and Level 5 driving automation. For the time being, it appears that US and German companies[28] are leading in this race. The US leadership is due to the efforts of both automotive startups and incumbents, whereas the German position is due to the efforts of incumbents.

Achieving Level 5 driving automation that will enable vehicles to operate under "easy cases" (e.g., highway driving), as well as in "difficult cases" (e.g., urban driving), without a driver being available to take over, will require:

- Several new technology innovations, in addition to the many that have already been introduced;
- Extensive additional testing under many different conditions so that regulators and consumers can gain the necessary trust in what will be essentially robocars.

Key technologies enabling autonomous and driverless vehicles

Autonomous and driverless vehicles are robots on wheels. These vehicles have large computing and data storage requirements. Even though we continue to miniaturize computing and data storage components while increasing their power, and have made great progress in these areas since the days of DARPA's Grand Challenge competitions, we still cannot accommodate all the necessary resources on-board such vehicles. Today we need to continue dividing the computing and data storage resources that make autonomous driving possible between the vehicle and the cloud (private or public). The field of cloud robotics[29] provides some of the necessary solutions.

Bringing to market vehicles with increasingly higher levels of driving automation requires technology innovations in four areas:

1. **Hardware technology.** Continued progress must be made in sensors, actuators, and processors, including sensor size, resolution, accuracy,

energy consumption, and price. Today the smartphones we carry have better sensors than most new vehicles. The opposite will be the case in the next few years when vehicles with Level 4 and Level 5 driving automation will be equipped with thousands of sensors. Work must also continue in actuators and processors to increase their speed while reducing energy consumption. For example, though Lidar sensors are being used extensively by autonomous vehicle prototypes as one of the approaches for sensing the environment where the vehicle operates, it is well documented that this sensor type is still too expensive for all by high-end vehicles and not as effective in adverse weather conditions.

2. **Software technology.** Software technologies such as the vehicle's operating system and the applications running in and around the vehicle, particularly the machine intelligence applications that exploit big data, are key enablers of ACE vehicles.

First, the operating system governing an ACE vehicle is far more complicated than the corresponding system in conventional vehicles and more crucial for the vehicle's ultimate acceptance by the public. It has to be secure, complete, and fail-safe in order to be trusted. Moreover, because of these requirements, the operating system's complexity increases as one moves from Level 1 to Level 5 driving automation.

Early examples[30] of ACE vehicles clearly demonstrate that their operating systems are highly configurable, and constantly updated software platforms that will run on new forms of infrastructure and application software. Parts of this software will be delivered as a service much like every other enterprise and consumer application software is today. Next-generation automotive operating systems enable new features and capabilities to be introduced on a continuous basis rather than on a model-year basis. As a result of the operating systems, and other software technologies that will be incorporated in ACE vehicles, we are starting to see the emergence of a completely new ecosystem of suppliers and OEMs that operate under new business models. For example, today Tesla offers drivers the ability to activate new features on the fly[31] and test them for a limited time. The capabilities of these new operating systems will necessitate that the automotive industry rethink the existing value chain, including in areas such as safety regulations, actuarial considerations, and financial underwriting considerations, as well as cybersecurity[32] and data privacy.

Second, because of the complex and unique user experience in vehicles with increasing levels of driving automation, the application software running in such vehicles, as well as relevant application software running on the mobile devices used by the occupants of such vehicles, plays a more critical role than what we have seen before in conventional vehicles. The application software running in such vehicles must be intuitive to use. It must perform information fusion, aggregation, and presentation, enable productivity and collaboration, and make possible a rich set of streaming and vehicle-based entertainment features that are accessible by the vehicle's occupants while being transported. Machine intelligence applications are a particularly important class of applications that is used for the exploitation of the big data produced by, within, and around electrified autonomous vehicles (with driving automation Levels

1–5), and by on-demand Mobility Services. This topic will be the focus of Chapter 4.0.

It is fair to say that to date the automotive industry has *not* developed, tested, and broadly deployed software of such complexity that simultaneously requires the levels of security, reliability and trust the operating system and application software of ACE vehicles will require. It is expected that the characteristics of this software will surpass the software currently used in state-of-the-art commercial airplanes. If incumbent automakers are to play a leadership role in electrified autonomous vehicles, not only they will need to improve their ability to develop and deploy such operating system and application software, but they will also have to gain the public's trust[33] that the software they produce is of the high quality and reliability autonomous vehicles require.

3. **Mapping**. While we talk about the sophisticated hardware and software in state-of-the-art autonomous vehicles, *another enabling component* for making the driverless future a reality is the digital map. Digital mapping is different than cartography. It is about the digital representation of the physical world. At a minimum, the digital map, in combination with big data about the environment where the vehicle operates and machine intelligence software, enables the autonomous vehicle to operate safely and successfully navigate to the intended destination. This is why digital mapping is used by vehicles with various levels of driving automation and by companies offering mobility services, such as ridesharing.

Today we are using mapping applications on our smartphones daily. In fact Google Maps is the most popular smartphone application in the world. Today's digital maps, while sufficient for personal use on our smartphones and for use in vehicles with lower levels of driving automation (Level 1 to Level 3), are necessary *but not sufficient* for driverless vehicles. They don't provide enough **resolution**, they are not as **accurate** as is necessary, and for the most part are **available** only **online**. Driverless vehicles require *high-definition maps* for the machine intelligence software to work properly. The resolution of conventional digital maps is at the road level. The GPS used for such maps provides meter-resolution. The resolution of the high-definition map needs to be at the lane level, i.e., it needs to be at centimeter-resolution. The high-definition maps need to also be tracking the various objects that impact driving, e.g., trees, road signs, bridges, etc., and the relations among them, e.g., the relation between the curb and the road. This is necessary in order for the vehicle to understand its position in the environment where it operates. For example, newer cars with Level 2 and Level 3 driving automation are equipped with a feature to keep them in a lane of traffic. For this feature to work properly, the existing digital maps are sufficient. However, in a multi-lane road, the driverless vehicle will need to know not only that it should stay in a particular lane but in which lane it is so that it can determine how to safely traverse lanes and how to proceed in a freeway exit.

Most digital maps today are updated at best every six months[34] and in many instances far less frequently. There are several reasons for this timing. Autonomous and driverless vehicles will require a more frequent updating of the digital maps. For example, the vehicle must be aware of construction projects, road closures, changes in road signs, changes in the location of venues, etc., all of which can impact the vehicle's ability to operate and transport its passengers to the desired location. Ideally, such updating must occur on a daily basis. Of course, the more accurate the map, the more likely it is that the high-frequency updating would become an impossible goal. For this reason, we will need to invent new ways for obtaining such updates, including a) crowdsourcing (as HERE will soon start to offer)[35] b) collaborating with startups like Nauto[36] and Comma.ai[37] that are capturing relevant data, and c) finding ways to capture vehicle-to-vehicle and infrastructure-to-vehicle communication data.

Finally, accurate high-definition maps must also be available offline so that the autonomous vehicle can continue operating safely under any condition. This is because a) certain types of sensors such as cameras and Lidar do not work appropriately in inclement weather or provide only "limited visibility", and b) the vehicle may need to drive through locations with poor broadband connectivity.

Today five companies are in a race to create high-definition digital maps because of the role they want to play in the driverless future: Google, HERE, Apple, Uber, and TomTom. Each of these companies realizes the importance of having their own digital mapping platform and has the resources to develop it. Ridesharing companies, like Uber, have different mapping needs than OEMs like Ford or Tesla. Ridesharing companies need to map only specific urban areas where they operate. Automotive OEMs must offer high-definition maps for every area where their vehicles are expected to operate, e.g., the continental United States. The investment to create such maps is very high and acts as an important barrier to entry for additional competitors. For example, Uber is investing $500M[38] on mapping, after failing to acquire HERE.[39] It is also acquiring mapping companies.[40] HERE itself has deployed several hundred special vehicles[41] collecting data in order to develop such maps. HERE is also discussing investments by Amazon, Microsoft[42] in order to continue spending in this area. Google continues to acquire mapping companies,[43] and so is Apple.[44] Tesla is starting to take a different approach by relying on the data collected by the cars it has sold rather than specialized vehicles.[45] This is an approach that can also be taken by incumbent automakers. By equipping with the appropriate sensors even a small percent of the millions of vehicles they produce annually, they will be able to efficiently collect data that can be used in high-definition digital maps.

High-definition maps are big data. Each of HERE's special vehicles collects 1TB of data per day.[46] In addition to data collected from these vehicles, high-definition maps must incorporate data[47] collected from satellites, other airborne sensors such as drones and airplanes, as well as from mobile devices such as smartphones.

4. **Broadband connectivity.** Today many vehicles are factory-equipped with 3G cellular modems that provide sufficient bandwidth for infotainment, including certain navigation applications. Analyst firms such as Frost & Sullivan estimate that 18 million connected vehicles are on the road today. This number will increase to 50 million by 2025. However, the bandwidth enabled by 3G technology, or even the bandwidth enabled by 4G or even 5G technology, may not prove sufficient for the data volumes envisaged in typical Level 4

and Level 5 autonomous vehicle operating scenarios. The data volumes generated by such vehicles, and the amount of communication between vehicles, and between vehicle and infrastructure, such as roads, toll collection, etc., is enormous. Autonomous vehicles need to communicate constantly with other such vehicles in order to make their presence known, and *share intelligence*, e.g., the existence of a newly discovered obstacle on a particular road. Imagine what happens when thousands of such vehicles are simultaneously on the road. Startups are already working with telcos and automakers to devise communication strategies and new technological approaches that combine cellular with other forms of broadband communication technology, e.g., Wi-Fi, to address the autonomous vehicle's big data communication needs.

3.3 BUSINESS MODEL INNOVATIONS: ON-DEMAND MOBILITY SERVICES

Experienced entrepreneurs know that **technology innovation** must be coupled with **business model innovation** in order to lead to disruption. To be successful ultimately these two types of innovation must feed off each other. This is also the case in the automotive industry. Already technology innovations in software, big data, including mapping data, hardware, and connectivity, combined with novel business models are leading to disruptive solutions that are changing the vehicle characteristics and the nature of personal transportation. But while we are often in awe of the technological innovations found in new and future vehicles, we have not paid as much attention to the business model innovations that are enabled by technology but are not themselves technological. This is because for the longest time vehicle ownership was the de facto model.

Vehicle ownership involves a limited set of well-understood business models. One could buy a vehicle or lease it under a relatively long-term contract. Car rental was considered only as part of long-distance travel, even when only a small amount of driving was involved in such travel, and taxis were the only viable alternative to mass transit in cities.

The connected vehicle started to enable new business models in many industries that work around the automotive industry. For example, states are starting to experiment with taxing consumers by the mile traveled, insurance companies are starting to use usage-based policies, the entertainment industry has adopted streaming radio and video subscription models, and the advertising industry is employing online advertising on vehicle maps[48] to monetize services and content.

However, the arrival and fast consumer acceptance of on-demand Mobility Services, enabled by software, big data, and smartphone technologies, is emerging as the real *business model disruptor* of the automotive industry. On-demand Mobility Services enable consumers to have access to vehicles without owning them, offer personalized transportation solutions and, more broadly, shape the future of mobility while addressing all four of the societal and urban challenges mentioned before.

Of all the different types of Vehicle Use services shown in Figure 4, we believe that the biggest impact will come from:

1. Ridesharing services, particularly those offered by Transportation Network Companies (TNCs) e.g., Uber,[49] Lyft,[50] and Didi Chuxing[51] to name but the three top companies in this sector.
2. Carsharing services,[52] e.g., Zipcar and Car2Go.
3. Parking services, e.g., Luxe, ParkMe, JustPark

4. Multimodal transportation enabling services.[53] The technology-driven management of multimodal transportation will itself enable a series of new business models. Studies from various management consultancies such as PwC[54] and Deloitte[55] are already discussing the management of multimodal transportation. For example, recently, Uber announced a partnership with local transit application Moovit,[56] to allow passengers of public transportation systems to seamlessly go from their originating point to a place where they can catch public transportation, and from where public transportation drops them off to their final destination.

These and other Mobility Services[57] that are transforming and disrupting the automotive and transportation industries demonstrate that, while they are enabled by technology, at their core one finds business model innovation. For example, the dynamic pricing (surge pricing) business model offered by ridesharing services is enabled by technologies such as location-based services software and GPS, the sophisticated analysis of traffic data, analysis of consumer demand data, and analysis of driver-supplied data. Similarly, car sharing service companies like Zipcar are testing a business model to charge drivers per mile driven rather than per hour.

The per mile rates that will be offered by TNCs adopting driverless vehicles means that in many cases, particularly in urban environments where people usually drive less than 3000–6000 miles per year, ridesharing will prove to be significantly cheaper than owning a car. If we succeed in convincing consumers to forgo owning a car, or at least to reduce the number of private cars they own, and use instead on-demand Mobility Services and/or multi-modal transportation solutions to address their transportation needs, we will have the potential of limiting the use of urban land for transportation infrastructure and increasing its use for other purposes. In addition to having the potential to radically impact urban planning and design,[58] this trend could mean fewer cars on the road, fewer accidents, and lower insurance costs for the consumer.

3.4 THE BIG SHIFT: FROM CAR OWNERSHIP TO CAR ACCESS

Venture investors see a recurring pattern in companies that disrupt markets or companies that create new large markets. These companies **identify** a problem that needs to be solved, and **postulate** a *big idea* that can result in the solution of the problem. If adopted, the solution leads to a **big shift** that typically results in **disruption of incumbents**, or the creation of the new market. Over the next 30–40 years ACE vehicles coupled with on-demand Mobility Services and associated multimodal transportation have the potential to provide solutions that combine technology with business model innovations and will address the problems discussed previously leading to a big shift in the car usage model.

For generations, owning a car has been a primary consumer aspiration in developed and developing economies. In such economies, the car is placed at the center of every person's life, and we often use it to define our self and net worth. Affluence is demonstrated first by owning a car and later by having bigger, faster, more luxurious, less-fuel-efficient vehicles (until Teslas arrived), as well as having more vehicles per household. Economic expansions have been reported in terms of increased car-buying activity and recessions by reversals in such activity. However, while car ownership will remain dominant and even continue to increase in certain developing economies, e.g., Southeast Asia, in certain economies, and particularly in urban settings, we have started to see a Big Shift. Driven by one or more of the problems discussed earlier in the chapter, consumers, particularly younger ones, are transitioning from the notion that puts car ownership at the center to one that puts car access[59] at the center. In these cases access is provided by Mobility Services that offer on-demand use of vehicles and broader multimodal transportation solutions. With growing consumer interest and the active involvement of transportation authorities, we expect that the range of such Mobility Services will increase, and their adoption will accelerate as ACE vehicles become available. In fact, because of the impact driverless mobility will have to their operating costs, fleet management companies, including Transportation Network Companies offering ridesharing services will most likely adopt driverless vehicles faster than individuals. Recent surveys conducted by academic institutions and management consulting firms show that globally consumers are interested in driverless vehicles, with younger consumers being more accepting than older ones.

In addition to driving, consumers are re-thinking many other aspects of car ownership: financing, insuring, parking, and servicing a vehicle. Based on surveys conducted by Arthur D. Little, the separation between car sharing, renting, leasing, and owning for both consumer and corporate vehicles is diminishing in the eyes of the consumer. According to these surveys the car is starting to be viewed as only one of the means that can move us through our daily life rather than something that defines us.

The big shift will not happen overnight and car ownership will not be abandoned. As with other big shifts, such as brick and mortar retailing to ecommerce, it may take a few decades. We expect that we will ultimately reach a *hybrid model that combines car ownership with on-demand car access*. In certain settings, e.g., rural areas, car ownership will remain dominant, whereas in others, e.g., cities and particularly megacities, car access will become the dominant form of transportation, particularly for certain population segments. We anticipate that the resulting hybrid model will imply fewer cars on the road with all the benefits this entails. On-demand car access of vehicles with guaranteed always-on broadband connectivity will improve productivity, as passengers are able to use their travel time more effectively. An illustrative early example of this case can be seen today in the Internet-equipped buses being used by the employees of high-tech companies in Silicon Valley,[60] and the increasing coordination among Silicon Valley's various transportation organizations.[61]

The hybrid model *will* impact automakers, their supply chain partners, and the existing business models[62] though it is not yet clear *how broadly*. It may impact each automaker differently depending on the market segment they focus on, e.g., luxury vehicles versus low-end commodity vehicles. For example, affluent consumers living in urban areas of developed economies may forgo owning a second luxury vehicle in favor of using ridesharing. On the other hand, a middle class family living in a suburb may continue to own or lease a mid-size car and a minivan.

4.0
BIG DATA AND MACHINE INTELLIGENCE IN AUTONOMOUS VEHICLES

A lot of attention is paid to the hardware, operating system software, certain types of application software, and connectivity technologies enabling the ACE vehicle. With the exception of using neural networks to create vehicle navigation models relatively little attention is being paid particularly by automotive industry incumbents on how to broadly exploit Big Data[63] in ACE vehicles and Mobility Services and use it as an asymmetric, long-term, and value-creating advantage. Transportation Network Companies are already demonstrating how the application of machine intelligence on the big data they collect can be used to offer a personalized customer experience, predict demand, dynamically set and optimize prices, and perform location-based analytics. These companies are actively working to expand the use of such technologies on additional applications, including driverless vehicles. As we shift to a driverless future in a hybrid model that combines car ownership with car access, Big Data and its exploitation using machine intelligence represent a unique, and currently overlooked, opportunity for automotive incumbents to lead in next-generation mobility.

The autonomous connected vehicle, i.e., the vehicle with Level 3, Level 4, or Level 5 driving automation, can be a big data and machine intelligence[64] hub. Its navigation system alone can be considered an insightful application.[65] As an insightful application, it:

1. **Generates** big data from its sensors, e.g., Lidar, cameras, the GPS, its actuators, the engine, its infotainment system, etc.;

2. **Receives** big data from the environment where it operates, e.g., intelligent infrastructure,[66] other vehicles, etc.;

3. **Analyzes** the generated and received data through a variety of models and makes predictions based on the goal it is trying to achieve, such as reach a specific destination. For example, it predicts that in order to reach its destination, it must turn in 100 ft. from its current position;

4. **Formulates** action plans that are based on these predictions. Each plan consists of one or more actions. For example, based on the prediction that it must turn in 100 ft., it formulates an action plan that could consist of actions including:
   - **Step 1:** activate the turn signal when the vehicle is 100 ft. away from the turn,
   - **Step 2:** come to a complete stop because there is a Stop sign at the end of the street,
   - **Step 3:** apply the brakes to start reducing speed, etc.;

5. **Applies** the action plan, e.g., execute Step 1, followed by Step 2, etc.;

6. **Evaluates** the results of each action.

And the cycle begins again, *millions of times* in the course of a single trip. Obviously, during each such cycle, new big data is generated.

Autonomous vehicles utilize machine intelligence in order to analyze data, formulate action plans, apply action plans, and evaluate their results, i.e., to make navigation possible, and control the vehicle. They can also use machine intelligence to provide a unique and differentiated user experience inside and outside the vehicle. The higher the level of driving automation, the more complex and multifaceted the machine intelligence that will need to be employed to make autonomy possible and the in-vehicle user experience seamless and natural for the driving environment. In-vehicle user experience is not about having a bigger monitor on the vehicle's dashboard. It is about considering the *entire transportation experience* and always providing the right information, at the right level of detail and through the appropriate modality, e.g., text, voice, video, for the ACE vehicle's passengers to comprehend and appreciate. Personalized on-demand Mobility Services require yet another layer of machine intelligence in order to provide a user experience that is commensurate to the vehicle's sophistication.

4.1 ACE VEHICLES AND MOBILITY SERVICES GENERATE BIG DATA

Next-generation mobility would require ACE vehicles to collect, process, manage, and analyze massive amounts of data. This data is:

- Generated by the vehicle, as in the case of telematics and vehicle-borne sensors (e.g., engine sensors, electrical system sensors, tire sensors, suspension, steering, radar/Lidar for longer-range sensing, and cameras for sensing in shorter distances). For example, Tesla vehicles today capture data about a) the controls the driver is using at any point in time while operating the vehicle, b) the location of the vehicle at any point in time, c) remote services accessed by the vehicle (such as remote lock and unlock), and d) the systems that impact the vehicle's operation, among others. It is expected that ACE vehicles will generate data in excess of 1GB/sec[67] during regular operation.

- Generated by and exchanged with other autonomous vehicles as they try to coordinate with one another and with the transportation infrastructure to ensure passenger safety. For example, the recent fatal accident involving a Tesla Model S running on Autopilot[68] and a truck operated by its driver may have been avoidable had the two vehicles been able to exchange data and automatically analyze and act upon this data.

- Generated by third parties, such as mapping data services.

- Generated by the infrastructure, e.g., roads, bridges, traffic lights,[69] toll-collection stations, etc.

- Collected through the applications of Vehicle Use services such as those shown in Figure 4, personal productivity services, and other services (traffic data, software updates, maintenance service records, extended warranty records, insurance records, personal calendars, social graph data).

Of the companies offering Vehicle Use services, the big data collected by parking, ridesharing, and carsharing companies is particularly important. In the course of conducting business on a global scale, ridesharing and carsharing companies capture, analyze, and exploit:

1. *Passenger* profile and preferences, including credit card data, through which they can access a variety of financial and demographic information, as well as behaviors in different contexts, e.g., Passenger A prefers a black car for rides to the airport (travel-related), prefers

multi-passenger rides to music concerts (entertainment-related), and prefers regular car rides to meetings (business-related).

2. *Driver* characteristics and behaviors in different contexts, e.g., Driver B belongs to the top 1% of drivers in terms of service provided during trips of 30 miles or more, based on passenger reviews and ratings, and low accident reports.

3. *Vehicle* data, e.g., vehicle breakdowns by make, model, and year based on reported incidents.

4. *Geolocation* data, including the starting point and destination of each trip, which they use in order to establish each ride's price and thus remain competitive with a passenger's other transportation options.

5. *Traffic* data, including road conditions due to repairs and accidents, which they use in order to more accurately estimate the time of the car's arrival, offer routing instructions to drivers in order to shorten the ride, and provide a better overall experience to the passenger.

6. *Site condition* data, e.g., airport construction projects, parking availability, conditions around event venues, e.g., stadiums, concert arenas, in order to improve the passenger experience.

As they experiment with new Mobility Services, such as on-demand goods shipping and delivery, ridesharing companies will collect additional types of consumer data, e.g., preferred grocery chains that can be used to further improve the user experience.

4.2 A FRAMEWORK FOR ORGANIZING MOBILITY BIG DATA

The various types of collected big data can be organized into a framework, shown in Figure 6. The framework initially organizes the big data across three dimensions:

1. *Data type*, depending on whether the data pertains to the vehicle or the driver/passenger(s)

2. *Data storage location*, depending on whether the data is stored in the cloud (public or private) or on-board the vehicle. We should note that because of the different types of big data that will be collected, cloud-based storage will likely involve many different public and private clouds. Each automotive OEM and Tier 1 supplier will need to have one of more data clouds (some regional, others central). In addition, each Mobility Services vendor will own similar data cloud configurations. Vendors offering mobility applications may use private clouds or public clouds such as Amazon's AWS to store the data they collect. Finally, we expect the emergence of cloud services dedicated to the management of transportation-related big data.

3. *Data half-life*, depending on how quickly the data is used, or needed, after it is captured, and for how long it is useful after it is captured, represents another important consideration in determining where the data is stored. For example, data about a road construction project can be captured and incorporated in the mapping system, and utilized, as soon as it is generated, for the autonomous vehicle's navigation. But such data also has longer half-life because it can also be used to refine and improve the predictive models that are part of the vehicle's navigation system. Other data, such as the traffic conditions around an entertainment event is useful only during the event and for short periods before and after the event.

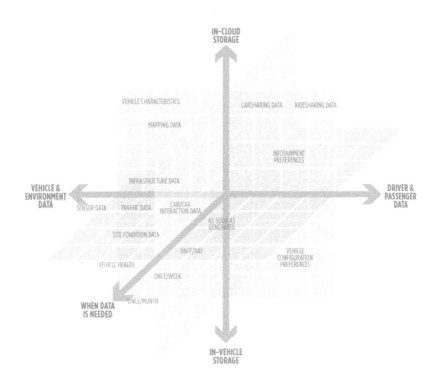

FIGURE 6: DIMENSIONS FOR ORGANIZING VEHICLE- AND MOBILITY SERVICES-RELATED DATA. FIGURE COURTESY OF EVANGELOS SIMOUDIS

This framework can be expanded to include additional dimensions such as degree of difficulty of obtaining each type of data, e.g., data provided by a smartphone application vs data that requires a special sensor.

The producers of such big data, in collaboration with its consumers must agree on the following:

1. What data needs to be aggregated and summarized over time, and what data needs to remain elemental?

2. What data needs to be communicated immediately, and what needs to be stored on-board the vehicle for later processing?

3. What data must be processed on-board the vehicle, what data needs to be processed in the cloud but in the "vicinity," e.g., geography-specific processing, and what data can be processed in the cloud in a central data center?

4. What data needs to be communicated between the various cloud services being used, i.e., cloud to cloud?

5. What big data needs to come into the vehicle from one of the clouds being used, other vehicles, or the transportation infrastructure in order to be used as-is, or combined with other data that is collected by various sensors?

To better understand how to use this framework to organize and manage the big data, it is necessary to identify and understand each of the applications that will exploit it. Figure 7 shows an example application intended to be used by consumers, in this case a ridesharing application, and the data sources it utilizes.

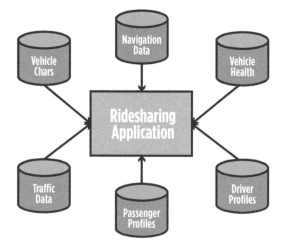

FIGURE 7: AN EXAMPLE OF THE DATA USED BY THE RIDESHARING APPLICATION. FIGURE COURTESY OF EVANGELOS SIMOUDIS

Several types of big data applications can be developed. Applications that address many aspects of the more traditional vehicle planning, manufacturing, distribution and sale part of the automotive value chain, as well as applications for the Vehicle Use part of the value chain. Some of these, like warranty claims analysis, that have been developed in the past using statistical analysis can be re-imagined using big data and machine intelligence. Others, like power management and range prediction for electric vehicles, or risk analysis for usage-based insurance will be developed for the first time. A comprehensive list of big data applications that have already been developed, are now being developed, or will need to be developed for the entire automotive value chain, including various types of Mobility Services, is shown in Figure 8.

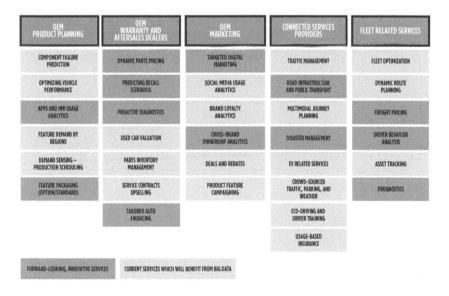

OEM PRODUCT PLANNING	OEM WARRANTY AND AFTERSALES DEALERS	OEM MARKETING	CONNECTED SERVICES PROVIDERS	FLEET RELATED SERVICES
COMPONENT FAILURE PREDICTION	DYNAMIC PARTS PRICING	TARGETED DIGITAL MARKETING	TRAFFIC MANAGEMENT	FLEET OPTIMIZATION
OPTIMIZING VEHICLE PERFORMANCE	PREDICTING RECALL SCENARIOS	SOCIAL MEDIA USAGE ANALYTICS	ROAD INFRASTRUCTURE AND PUBLIC TRANSPORT	DYNAMIC ROUTE PLANNING
APPS AND HMI USAGE ANALYTICS	PROACTIVE DIAGNOSTICS	BRAND LOYALTY ANALYTICS	MULTIMODAL JOURNEY PLANNING	FREIGHT PRICING
FEATURE DEMAND BY REGIONS	USED CAR VALUATION	CROSS-BRAND OWNERSHIP ANALYTICS	DISASTER MANAGEMENT	DRIVER BEHAVIOR ANALYSIS
DEMAND SENSING – PRODUCTION SCHEDULING	PARTS INVENTORY MANAGEMENT	DEALS AND REBATES	EV RELATED SERVICES	ASSET TRACKING
FEATURE PACKAGING (OPTION/STANDARD)	SERVICE CONTRACTS UPSELLING	PRODUCT FEATURE CAMPAIGNING	CROWD-SOURCED TRAFFIC, PARKING, AND WEATHER	PROGNOSTICS
	TAILORED AUTO FINANCING		ECO-DRIVING AND DRIVER TRAINING	
			USAGE-BASED INSURANCE	

FORWARD-LOOKING, INNOVATIVE SERVICES   CURRENT SERVICES WHICH WILL BENEFIT FROM BIG DATA

FIGURE 8: BIG DATA APPLICATIONS IN THE AUTOMOTIVE VALUE CHAIN. FIGURE COURTESY OF FROST & SULLIVAN.

Three additional application types are missing from the list provided in Figure 8. First, **personalization applications** for configuring the transportation experience both inside the vehicle and for providing customized on-demand Mobility Services. For example, an application providing parking options around entertainment venues on days when an event is taking place. Second, **mapping applications** that are implemented on top of the next-generation high-definition mapping platforms described in Chapter 3.2. For example, Tesla uses a big data application that analyzes vehicle data, traffic data, mapping data, and owner preference data to determine a) the number of re-charging stops a vehicle must make during a long-distance trip, and b) where to make each of these stops in order to provide driver and passengers with the best customer experience. In making these determinations the application balances number of stops with amount of time spent during each stop. Finally, **urban planning applications**.[70] For example, Mobile Market Monitor has developed an application that informs a city's transportation planners of the transportation modes each consumer uses during a day.

4.3 BIG DATA EXPLOITATION USING MACHINE INTELLIGENCE

Machine intelligence is key for the exploitation of big data in autonomous vehicles and next-generation mobility solutions. I use the term "machine intelligence" rather than the term "machine learning" because it is broader and includes a variety of artificial intelligence approaches, such as computer vision, speech recognition and synthesis, natural language processing, machine learning and deep learning, that can be applied to exploit the big data generated in, by, and around autonomous and driverless vehicles, and by on-demand Mobility Services. However, within the range of machine intelligence techniques, of particular importance is **machine learning** and specifically **deep learning networks** because of the types of big data generated by ACE vehicles and on-demand Mobility Services and their effectiveness in vehicle navigation and several of the other applications mentioned in the previous section.

Machine learning is a part of Artificial Intelligence (AI)[71] (a higher-level introduction to AI is provided in this article).[72] It refers to the computer programs and methods that enable computers to acquire new capabilities without being explicitly programmed but only by being exposed to new data. There are many approaches to machine learning.[73] For the navigation functions of autonomous and driverless vehicles, deep learning networks,[74] a branch of machine learning that represents an evolution of artificial neural networks,[75] are proving to be the most successful. This is because the autonomous vehicle has to *perceive* and *respond* to complex situations, e.g., multiple moving objects, such as other vehicles, pedestrians, etc., under varying light, weather, and road conditions, and deep learning networks are extremely good at these complex situations. Deep learning provides complex representations of data while making computers independent of human knowledge. It extracts representations directly from unsupervised data without human interference. These networks learn, or are trained, by being presented millions of different examples of each specific situation, for example, of cars driving in a multi-lane freeway. Each example may be represented by millions of features such as those that are extracted from a picture or a video frame. As a result, training using deep learning networks may take hours, or even days, depending on the concept that is being learned. The training of such networks to create navigation models requires vehicle data, data about the environment where the vehicle operates, and mapping data. Over time, through the examples they are being presented, deep learning networks are able to automatically form abstract representations of complex situations. They can then apply these representations on new instances of similar complex situations and make an appropriate prediction, e.g., turn in 100 feet.

Companies like Google and more recently Tesla[76] and Uber[77] have been collecting live data from the sensors of the vehicles they operate. However, even these big data sets often prove inadequate for the proper training of deep learning networks used for vehicle navigation. As a result, these companies are starting to also generate simulated big data[78] in order to properly train the deep learning networks they use for the navigation applications of their autonomous vehicles.

In addition to enabling the driverless vehicle to navigate autonomously, machine intelligence, including other types of machine learning, can be used in order to develop big data applications that are able to:

- **Predict** the behavior of other vehicles in the vicinity of a driverless vehicle, as well as that of pedestrians. This is a particularly hard problem, because even with the existence of driving rules, there are driving conventions that are often regional and override the rules, e.g., the characteristics of "Boston driving conventions." Certain such behaviors may be deeply personal. For example, a driver may prefer to drive aggressively, or a pedestrian may prefer to jaywalk.
- **Create** a complete understanding of the vehicle and its passengers, enabling a tailored in-vehicle user experience and the dynamic optimization of the vehicle's performance.

- **Improve** the vehicle's usage economics, which will be particularly important to Transportation Network Companies and other types of fleet management companies.
- **Provide** personalized transportation experiences and solutions to driver and passenger, including personalized experiences at the destination. If the car becomes just one of the means for moving through daily life, then passenger and driver would want the next-generation vehicle to be able to take into account their needs not only prior to the beginning of each trip but also at the destination (e.g., parking, safety, dining options), their broader transportation services, (e.g., public transportation options and schedule, ridesharing, carsharing, car rental), and perhaps even some personal context about a passenger prior to entering the vehicle and at the end of the trip. (For example, if I was running on my way to my driverless vehicle, something which my mobile phone or fitness device is able to detect, then the vehicle's ventilation system should set to an appropriate temperature; if I was listening to a particular music track prior to entering the vehicle, then the vehicle's infotainment system should pick that up). In this way, the vehicle provides driver and/or passenger with a continuous personalized experience.

The application of various machine intelligence approaches on the big data we have been discussing comes with serious technological[79] and ethical challenges[80] we will need to overcome. Though we are making great progress,[81] we are still in the early stages of understanding how to effectively, and safely apply these techniques on the data we are now collecting and the data we will be collecting during our driverless future, which will be orders-of-magnitude larger, faster, and more complex.

4.4 AN EXAMPLE OF MACHINE INTELLIGENCE IN ACTION

To demonstrate how automakers can gain a competitive advantage by providing value through the broad exploitation of big data produced in, around, and by the autonomous vehicle, let me present an example. This example showcases how machine intelligence can be utilized outside the navigation function in a scenario that combines car ownership with car access through Mobility Services. The goal here is not to discuss particular machine intelligence techniques but only to demonstrate how automakers can turn value-creation into a competitive advantage through the use of such techniques.

Most of the times when I go to San Francisco for business meetings, I drive my own car to the end of the freeway, park near the freeway, and use Uber to go to my meetings because of efficiency, convenience, and cost. On my commute to the city, I use Google Maps, rather than my car's navigation system because Google's routing, traffic information, and dynamic rerouting are better than the information provided by my car's infotainment system. As I approach the end of the freeway, I must decide where to park, since there is both street parking and parking garages available. Depending on the amount of time I will spend in the city I may first look for on-street parking and then use a mobile application to look up which of nearby garages have available space and their hourly rates. These are all seemingly simple but actually complex decisions that can be made automatically by analyzing big data using machine intelligence. They can greatly impact the user experience. Autonomous and driverless vehicles will have to make such decisions automatically.

Once I park, I pick up the first Uber ride in front of the garage and return there at the end of my business day. So, more or less, Uber knows my movements while I'm in San Francisco, along with all my transportation preferences. Through data analysis, Uber can even infer the duration of my meetings.

Imagine that the automaker of my car has a partnership with Uber. Further imagine that I opt in to allow my automaker to access my calendar and my Uber data. By combining mapping data, traffic data, parking data along with my personal data it now has permission to use, and applying machine intelligence the automaker can manage my mobility and offer me a personalized transportation solution for my business trips to San Francisco. Such a solution would include: a) the particular route to take on each trip to the city, b) the location of where to park, based on traffic at the time of the trip and parking space availability, e.g., I may need to park further outside the city on days when the traffic is very heavy or if there are no parking spots closer to the city, c) a discount offer for the recommended garage that is nearby to that location, and d) discounted Uber rides to my each of my city destinations in vehicles the automaker manufactures (by having access to my calendar, the automaker can schedule an Uber ride in a car of their brand, thus minimizing the wait and, most importantly, keeping my rides within their brand). By properly exploiting big data and taking all these actions, the automaker will provide me with a better customer experience.

Finally, by using big data and machine intelligence to understand my driving and transportation usage patterns, my car insurance carrier not only could proactively make me an appropriate Usage-Based Insurance offer but could change my risk profile and make me a better overall offer because of the way I use my car, i.e., I don't drive it in the city, where the risk of accidents, theft, and vandalism is generally higher.

The importance of big data and machine intelligence among all the other fundamental technologies that enable the next-generation mobility is becoming clearer by the day. Even though the big shift to a hybrid model that combines car ownership with car access will take decades to complete, starting to master these technologies now and using them to defend against disruption from new companies entering the mobility market and provide enduring value through a differentiated user experience, will be key for incumbent automakers. However, developing big data expertise will not come easily to the incumbents. First, they must accept to *be in the insights business, while they continue to excel in the vehicle design, manufacturing and distribution business.* This chapter has shown that in developing big data expertise incumbents must learn to:

- Create information from the big data they collect, e.g., understanding how driver and passengers use the vehicle.
- Utilize third-party data, e.g., weather data, mapping data, calendar data, to augment the value of the data they collect.
- Build insightful applications that exploit the available big data, e.g., a parking application that takes into account parking regulations, space availability, price, convenience and safety.

Second, they must undergo a radical transformation because *along with developing data expertise in order to provide superior transportation experience to their customers they must make other changes to the way they operate and make decisions.* These transformations will require incumbent automakers to adopt new innovation strategies to match the faster innovation cycles of information technology. They will also require changes to their corporate cultures. To better appreciate the characteristics of the innovation strategies and culture incumbents should adopt, it is beneficial to analyze the characteristics of four innovators that are at the core of next-generation mobility.

5.0
FOUR COMPANIES AT THE CORE OF AUTOMOTIVE DISRUPTION

hat started several years ago as DARPA's Grand Challenge, and continued in a few corporate research labs and more recently the garages of a small number of venture-backed startups, has now become a big wave that can disrupt the automotive and transportation industries. Our driverless future is about more than just the design, manufacturing and distribution of vehicles. It is about big data, information and information technologies, including sensors and semiconductors, it is about electrification and broadband connectivity, and it is about Mobility Services. While we now speak of hundreds of startups that are part of this disruptive wave, as the reader might have already surmised, I consider four companies as the key innovators and disruptors: Tesla, Zipcar, Uber, and Google. These companies and the areas of their core competence are shown in Figure 9. Note that Waze is now part of Google.

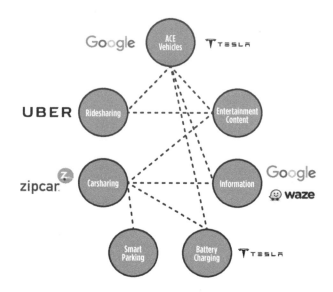

FIGURE 9: ACE VEHICLES AND THEIR INTERACTIONS WITH MOBILITY SERVICES DISRUPTING THE AUTOMOTIVE INDUSTRY. FIGURE COURTESY OF EVANGELOS SIMOUDIS.

- **Tesla.** Tesla is attempting to disrupt the entire automotive value chain, and like Apple it is attempting to control major parts of the value chain. The company's innovations started with the electric-connected vehicle[82] but also include several others,[83] such as charging stations, the car and battery manufacturing processes, as well as its direct-to-consumer sales and service model, personalized user experience inside and outside the vehicle, and automatic software updates. With the introduction of the Autopilot, Tesla's Model S and Model X vehicles are now equipped with Level 3 driving automation. Many of these innovations are driven by software and big data and are facilitated by Tesla's decision to control its entire supply chain,[84] much like Apple typically does for its products, and in a way that is also reminiscent of what automakers used to do in the past.

There should be no doubt that Tesla is disrupting by utilizing big data. For example, the telemetry being gathered from each car is used to analyze the entire fleet's usage patterns (that in turn is used to improve capabilities, such as the vehicle's battery range,[85] introduce new features), detect crashes, identify need for maintenance that can improve vehicle performance, and find lost cars. Just through the Autopilot feature, Tesla is collecting more than one million miles of driving data per day[86] and is now starting to use this data to create its own high-definition maps that are necessary for autonomous driving and driverless vehicles. In its roadmap for the future,[87] it has committed to offer Level 4 driving automation,[88] i.e., autonomous cars,[89] in the next few years. It has also stated its intention to offer a carsharing mobility service,[90] as well as a ridesharing service[91] as its vehicles become increasingly autonomous.

- **Zipcar.** Zipcar was the first to offer carsharing mobility services. Zipcar's innovations include its membership-based, carsharing disruptive business model that is combined with its big data software platform and novel software-driven user experience. Zipcar disrupted the car rental industry, and that's why Avis acquired the company.[92] As an independent division of Avis, Zipcar continues to analyze the big data it collects to: a) identify new locations to place cars, i.e., enabling it to establish a more distributed rental network and in this way offering more convenience to its members and increasing the probability they will use its service more frequently, b) better re-balance its fleet (fleet re-balancing based on usage is a big issue since one-way rentals represents 12% of North American carsharing membership), c) offer one-way rentals at more competitive prices than full-service companies, and d) offer lower prices/hour of usage and more recently experimenting with per-mile charging,[93] which will constitute another business model innovation.

- **Uber.** Uber, while not the first to offer ridesharing services, has become synonymous with the category because of the innovations it introduced, its global reach, and the capital it has raised. At the core, Uber is a big data company. In addition to innovating through its business model[94] (per mile pricing of individually-owned, rather than fleet, vehicles), it is also innovating with its mobile application that brings together software with big data, and the broad utilization of big data[95] in many facets of its operations such as demand-based dynamic pricing, and more recently autonomous driving. Its mobile application uses big data in the presentation of routing information, in offering transparency for the vehicle's arrival time, as well as in developing driver reputation. The company continues to expand globally with blinding speed as it aims to build barriers to entry, though it now faces extreme competition from Lyft and Didi Chuxing. After initially disrupting the taxi and limousine industries, i.e., a small part of its market potential, and becoming a company with more than $1B in annual revenue, Uber is now attempting to next disrupt the entire on-demand delivery industry and start competing with Amazon, Google, and logistics companies.

More recently Uber started work on an autonomous car project,[96] and, in partnership with Volvo, is testing autonomous vehicles in Pittsburgh.[97] Through autonomous vehicles Uber will attempt to better control its costs (cost/mile can go from \$1.6/mile[98] today to as low as \$0.31/mile[99] with an electric, autonomous car) enabling it to improve its margins and lower its prices in order to increase its ridership. To develop autonomous vehicle technology, Uber is using a multi-pronged approach. In addition to continuing to grow its internal big data team, it has partnered with Carnegie Mellon University (CMU), an academic leader in robotics research that fielded teams in each of DARPA's Autonomous Vehicle Grand Challenge competitions, opening a facility in Pittsburgh[100] that is staffed by several CMU researchers it has hired. It acquired Microsoft's mapping assets and know-how[101] and is making a large investment to develop high-definition maps.[102] Finally, it has recently acquired Otto,[103] a 90-person startup that has been developing self-driving truck technology. This acquisition significantly expands Uber's autonomous vehicles team in addition to providing the company with additional know-how in relevant technologies. We should note that Uber is not alone in thinking along these lines. Lyft, one of its biggest competitors, through its partnership with GM, and Zoox[104] a well-financed startup developing robo-taxis, are also working hard in this area.

- **Google**. Google is disrupting the automotive industry with two platforms. First, its Android Auto mobile platform (Android Auto is the name of the mobile platform) can control the car's dashboard, including the navigation platform that is based on Google Maps and incorporates high-definition maps. The data collected from this platform is combined with Google's data analysis capabilities to start providing an increasingly personalized in-vehicle experience, as well as an in-context experience when entering the vehicle.

Second, Google has been developing a self-driving car platform that can be used with Level 4 or Level 5 driving automation. Google started its self-driving vehicle experiments using modified Lexus SUV vehicles.[105] These vehicles can be driven autonomously or by a driver. More recently, Google also signed a partnership with FCA[106] to produce a small batch of Level 4 autonomous minivans. However, Google has also developed vehicles with Level 5 driving automation,[107] i.e., they are driverless. These vehicles will be used in a transportation network to offer mobility services. Google will have to partner with incumbent automakers or contract manufacturers, e.g., LG or Foxconn, that are starting to develop vehicle expertise, in order to create such a network of vehicles. Such vehicles will be running on Google's navigation platform, in a way that will be similar to the approach Google took with the Android operating system. By using big data analytics on data collected through this network, Google could develop applications that offer dynamic ride pricing to optimize the network's usage, optimize the number of vehicles that will be needed to serve a population, and other such applications. Google would very likely want to own the data generated by each vehicle and even have the exclusive right to monetize this platform through data-driven advertising.

Apple could emerge as a fifth major disruptor of the automotive industry. Today it is disrupting with its CarPlay platform that controls the car's dashboard, similar to Google's Android Auto. It is one of the five companies that own a mapping platform and could develop high-definitiion maps in the future. It is also developing a software platform for ACE vehicles. It is not clear yet whether it will produce its own ACE

vehicles or make its software platform available through vehicles manufactured by other OEMs or TNCs. Apple can disrupt the automotive
industry even if it chooses to focus exclusively on the user experience.
It could disrupt not only the car's software and hardware platforms but
also the overall car-buying experience, car-servicing experience, etc.
very much like it did with its mobile devices (iPod, iPhone, iPad).

5.1 THE CHARACTERISTICS OF THE DISRUPTORS

Though very different from one another, the four companies that have
been leading the disruption of the automotive industry share several
important characteristics that I consider as critical to their success:

- **Big data and machine intelligence companies, at their core**. As
  I have mentioned before, Tesla, Uber, Zipcar, and Google at their
  core are all big data companies that use machine intelligence to
  exploit the big data they collect and gain access to. Apple is becoming such a company by virtue of its mapping platform and mobile
  ecosystem. Google's ability to collect and exploit data is well documented and continues to evolve. Tesla has also been collecting,
  analyzing, and exploiting big data since day one, an activity that
  is now expanding with the introduction of the Autopilot. As I
  mentioned above, in the process of conducting business, Mobility
  Services companies also generate, collect, analyze, and exploit big
  data through machine intelligence approaches. As Uber and Zipcar
  are working to improve their service, introduce new services,
  reduce their operating costs, and better control their pricing, they
  are collecting and utilizing big data about the vehicles in their fleets
  and obviously their customers. Finally, Google, Tesla, Uber, and

Apple, by owning mapping platforms, they have developed, or are developing, important expertise in using this form of big data.

- **Direct-to-consumer model.** Another important common characteristic of these disruptors is their direct-to-consumer model. Through this model they have established a continuous dialog with their customers. Combined with their big data expertise, it enables them to offer a better and constantly improving user experience which by itself becomes a competitive advantage. Because of the dialog these disruptors receive immediate feedback to "experiments" they conduct about every aspect of their business: testing new products and product features, new sales models and new business models. For example, consider how Tesla recently enabled its customers to "unlock" the Autopilot feature on older Model S vehicles, or how Uber introduced music and food delivery through an update of its mobile application. By comparison, automotive industry incumbents are one or two steps removed from their consumer customers and currently receive the majority of their feedback from car dealers.

- **Venture-backed startups.** All four started as venture-backed startups with three of the four having their roots in Silicon Valley. Because of their founders and the venture capitalists they attracted as investors, they developed high-performance corporate cultures. They are extremely adept at identifying problem areas by understanding changes in technology, business models, and customer preferences. Uber was not the first to offer ridesharing services. However, it was the first to understand how to harness the technology capabilities of the smartphone, realize the consumer needs for a reasonably priced alternative to the taxis and limousines and offer it through a novel business model.

- **Experiment constantly.** A key characteristic of startups in general but venture-backed startups in particular is their ability to constantly experiment and iterate quickly until they find the right product/market fit. They do that before scaling their efforts. We have already mentioned that Uber started testing autonomous vehicles in Pittsburgh and that it is entering other areas of on-demand goods delivery.[108] Many of these experiments may lead to failures; see, for example, Uber's failing efforts to establish a significant business across Europe[109] and China,[110] and the human interventions needed in Google's autonomous cars.[111] Successful startups are able to acknowledge such failures, recoil, and adapt. In the process they are also able to constantly create new IP. They are also able to eliminate experiments[112] that will not lead to big successes.

- **Make decisions with incomplete data and uncertainty.** The biggest advantages startups have over large corporations is their agility and ability to pick up weak market signals. They combine these advantages with their ability to conduct their business with and make decisions from incomplete data and their willingness to operate with high levels of uncertainty, particularly when they enter new and unformed markets. Tesla didn't have data on whether consumers would embrace electric cars that cost more than $100,000 and had a range of 250 miles. Similarly, Uber didn't know whether consumers globally would be willing to abandon taxis and ride in the cars of strangers. And now Google doesn't know whether consumers will be willing to ride in driverless vehicles that don't even have a steering wheel. In all of these cases, these disruptors, using incomplete data and under uncertainty about the outcome, made "bet the farm" decisions.

- **Hire the best.** Startups in general and venture-backed startups in particular are known for always trying to hire the best people, keeping them focused on the right problems, and providing the right incentives that reward risk-taking and collaboration to keep them in place. Today there is an ongoing war for talent, particularly in Silicon Valley, Israel, and China, to staff the startups working on automotive startups. In addition to Uber's hiring of CMU's scientists to work on its autonomous vehicle project, Google is also aggressively hiring people for its self-driving car project.

- **Organized in small teams.** These disruptors have come to realize early on that it is important to not only hire the right talent but also to organize them in the right way. Small teams with flat structures, whose members come from diverse disciplines, have consistently outperformed the large highly hierarchical organizations often found in automotive corporations. Google, Tesla, and Uber all organize their employees in small, agile teams.

- **Take risks and break rules.** It goes without saying that working in a startup, even a well-funded venture-backed startup, is about taking risks. Every big decision is the equivalent of a moonshot. The four automotive industry disruptors took, and continue to take, big risks—some of which may prove fatal. For example, Tesla is risking large amounts of capital as it is trying to expand its manufacturing capacity while designing its next car (Model 3) because it is expecting high, ongoing demand for that vehicle. Uber continues to raise large sums of capital to fuel its global expansion while battling government regulation and the competition. Like every other disruptor, Uber, Tesla, and Google are known for breaking rules and worrying about the consequences later, thinking that, in many instances and in order to disrupt, you cannot wait for the appropriate laws and regulations to be created. You expect that the regulations will eventually catch up with the innovations that lead to the disruptions. For example, whereas incumbent automakers would have asked for regulator approval before testing a feature like the Autopilot, Tesla just downloaded the appropriate software on its cars and asked the drivers to opt-in.

At the turn of the 20th century, Detroit[113] had an extremely vibrant automotive startup ecosystem that numbered more than 120 companies. The area became an automotive innovation cluster with a large startup ecosystem because it had attracted a critical mass of entrepreneurial technology innovators and possessed natural advantages that suited automotive production (location close to production centers and to coal, iron, and copper production, accessibility via waterways), in the same way that today Silicon Valley offers advantages to the next-generation automotive disruptors because it is the center for software, big data, communications, semiconductor, and business model innovations. Ford, General Motors, and Chrysler, emerged out of that ecosystem and became leaders because of the innovations (technology, business model, employment) they introduced and the risks they took. To repeat a similar feat, these and other automotive industry incumbents will need to understand and appreciate the characteristics of the disruptors presented in this chapter, and determine how they need to change in order to create an enduring competitive advantage through big data in the driverless future of next-generation mobility. They need to consider new innovation strategies, one of which is presented in the next chapter.

6.0
INTRODUCING THE STARTUP-DRIVEN INNOVATION STRATEGY

Technology and business model innovations are leading to next-generation mobility solutions that are starting to impact, and can potentially disrupt, the automotive industry. We discussed the characteristics of the disruptors that are capitalizing on these innovations to leapfrog incumbents and establish leadership positions in the evolving transportation ecosystem. But why couldn't incumbent automakers appreciate sooner many of the over-the-horizon technological and business model innovations that are leading to on-demand mobility and a driverless future? Why were Google, Uber, Tesla, and several startups able to start working on innovative data-driven mobility solutions, and make relevant investments and acquisitions, before any of the automotive incumbents begun to focus on autonomous vehicles and on-demand Mobility Services? Why was it Google that acquired Waze,[114] a big data mapping application company, in 2013 rather than GM, Ford, or even HERE? Automotive industry incumbents, like corporations from many other industries such as logistics, consumer packaged goods, and financial services, are placed in this position because they routinely fail to create, or understand the potential impact of, over-the-horizon innovations and bring them to market under the appropriate business models in a way that will enable them to build next-generation businesses. Corporations like Google, Amazon, Facebook, and Apple are more successful at this task. These companies have startup roots and continue to behave like startups even as they mature. Additionally, when they invest in or acquire startups, they are more likely to turn them into successes with higher frequency compared to corporations that have lost their startup DNA.

As we analyze corporations we begin to see that many are particularly good at using their R&D organization to produce innovations that support, continue to scale, and even extend their existing business models. But they are rather weak at creating new businesses using innovative technologies and novel business models. While such businesses may generate new revenue streams, they may also cannibalize, or completely disrupt, a corporation's existing models. This possibility is particularly difficult for corporations to accept. In the automotive industry it has proven impossible for incumbents to predict that electric vehicles with many innovative characteristics will become popular even during periods of low gas prices, that consumers globally will be adopting ridesharing services with increasing frequency, and that new companies, including many startups, will be first to bring to market autonomous, and even driverless, cars with consumers and fleet management companies, such as Transportation Network Companies, ready to adopt them.

It is slowly starting to become apparent to corporations, including several from the automotive industry, that in the presence of this environment, the innovation model that is based solely on the efforts of corporate R&D organizations is no longer sufficient for addressing the disruption challenges they face and the long-term growth goals they need to achieve. In other words, the automotive industry must not only recognize the opportunity afforded by big data and machine intelligence, but also the need for a new innovation strategy that will enable it to take full advantage of this opportunity. For reasons explained here,[115] we are starting to understand that corporate R&D is losing its ability to effectively identify and adequately develop over-the-horizon innovations[116] that can protect the corporation from disruption or enable it to disrupt. Increasingly startups are succeeding in this role and in the process disrupting a variety of industries, even industries with high barriers to entry, such as automotive.

What we are witnessing with the role that startups play in the automotive industry is not new or unique to this industry. We have seen the pattern before in industries such as retail and telco: first, new technologies with new business models become better aligned with customer needs while the incumbents' R&D organizations are not able to respond; second, profit margins fall, and industry incumbents cannot adapt to changes; third, consolidation occurs and new leaders emerge.

To avoid this pattern and succeed in their transformation from manufacturing to information and transportation solution companies, corporations in the automotive industry need to reinvent their approach to innovation. Specifically, they need to:

1. Broaden their view on what constitutes innovation and consider *business model innovation* and other types of innovation, e.g., sales model innovation, to be as important and sought after as *technology innovation*;

2. Establish new ways to identifying and developing over-the-horizon innovations, including big data and machine intelligence innovations, that can become the next-generation businesses.

The new approach must involve and engage the startup ecosystem and must enable corporations to innovate:

- At a faster rate and more cost-effectively,
- From outside *and* inside by leveraging entrepreneurship and intrapreneurship,
- By combining technology with business model innovations,
- By establishing the right timelines for each type of innovation,
- By assessing the effectiveness of each innovation through a portfolio-management approach and Key Performance Indicators (KPIs) that are appropriate for the type of innovation and the corresponding timelines.

The **Startup-Driven Innovation Strategy** enables corporations to accomplish these goals.

6.1 DEFINING THE STARTUP-DRIVEN INNOVATION STRATEGY

The Startup-Driven Innovation Strategy is a corporate innovation strategy that describes how corporations can *innovate continuously* by a) *working with startups* and b) *applying approaches pioneered by startups and the members of their ecosystem*, e.g., their entrepreneurs, venture investors, etc.

To better understand the strategy, I have found it useful to build upon the work on the Three Horizons of Growth[117] framework. According to this framework, corporations must operate across three horizons. **Horizon 1 (H1)** includes the corporation's core businesses, the ones whose business models provide the greatest profits and define the corporation's brand. For example, the Tide detergent for P&G, the Series 3 compact car for BMW, or the 737 passenger jet for Boeing. **Horizon 2 (H2)** includes emerging opportunities that are either extensions of H1 businesses or growing new ventures that can eventually become H1 businesses. For example, the Boeing Business Jet represents an H2 business, as it is an *extension* of the 737 passenger jet. **Horizon 3 (H3)** includes the disruptive ideas that could provide profitable growth in the future, once the appropriate business model is identified. For example, IBM's Watson platform started as an H3 initiative based on work that was initially performed within IBM's R&D organization.

Most corporations focus their innovation efforts on H1 businesses, and some, such as Boeing, also invest in H2 initiatives. In fact, corporations such as Unilever[118] and American Express[119] have been visiting innovation clusters, such as Silicon Valley and Israel, and routinely partner with or acquire startups to gain access to specific technologies that could help them address their H1 innovation needs. We expect that this trend will expand, particularly as more corporations are establishing a permanent presence in these, and other, innovation clusters[120] with high startup concentrations.

Where corporations fail is in consistently identifying and launching H3 opportunities. The opportunity from big data and machine intelligence presented to the automotive industry because of the introduction of ACE vehicles combined with Mobility Services serves as an excellent example of the ramifications of such failures. To avoid such failures, corporations must *innovate continuously across all three horizons*, essentially becoming **3-Horizon Corporations**. And the Startup-Driven Innovation Strategy enables them to establish and manage effectively their innovation initiatives and become 3-Horizon Corporations.

Before embarking on applying the Startup-Driven Innovation Strategy, it is important for the corporation to determine the **type of innovations** it will pursue. Is it looking for H1 innovations to *maintain existing business models*, H2 innovations to *extend these models to adjacent markets*, or for H3 innovations that will *disrupt existing models and create new ones that may cannibalize existing business models*?

The Startup-Driven Innovation Strategy incorporates *organizational, financial,* and *cultural* elements.

- **Organizational**: The corporation must establish *Innovation Outposts*, whose mission is to enable the corporation to identify and respond to developments that can disrupt the corporation or allow the corporation to be disruptive. Innovation Outposts collaborate with the corporation's various existing business units and complement its R&D organization. Innovation Outposts are managed by a single executive leader who reports to the company's CEO.

- **Financial**: In addition to its R&D budget, the corporation must allocate sufficient capital for each Innovation Outpost's operation and the variety of initiatives it leads.

- **Cultural**: Innovation can come from anyone within the corporation, i.e., the corporation's intrapreneurs, as well as from outside the corporate four walls, particularly from startups and their entrepreneurs. The corporation must foster a *culture of continuous innovation*.

6.2 THE INNOVATION OUTPOST

The Innovation Outpost[12] is a corporate organization that operates in an innovation cluster. In addition to managing the overall Startup-Driven Innovation Strategy for the corporation, it **performs two basic functions:**

1. It **senses** for innovations that can serve one or more of the H1 or H2 businesses, become threats and disrupt the corporate parent, or can be used as opportunities to launch new H3 efforts.

2. It **responds** to the opportunities it identifies five ways.

   a. **Invent**: The Innovation Outpost can establish advanced development efforts that are associated with a specific innovation goal. It can also launch broader H3 basic-research efforts that take advantage of or investigate technologies and business models the innovation ecosystem is known for. For example, Verizon's Silicon Valley R&D center focuses on big data and software technologies, as well as online advertising-based business models.

   b. **Invest**: The corporation may allocate a venture fund to invest in startups working on technology and/or business model innovations of interest. For example, UPS invested in Ally Commerce in order to understand the logistics opportunities arising from manufacturers selling directly to consumers rather than through distributors.

   c. **Incubate**: The Innovation Outpost's incubator supports the efforts of early-stage teams and companies that want to experiment in areas that interest the corporation. For example, Samsung's incubator focuses on startups working on the Internet of Things. The teams may consist of entrepreneurs only, intrapreneurs only, or be mixed thus giving the opportunity to corporate employees to apprentice with entrepreneurs. The Innovation Outpost's incubator can also enable the corporation to experiment with new corporate cultures and work environments. For example, Standard Chartered Bank's startup studio was such an effort.

d. **Acquire**: Corporations buy startups in order to access both the innovations they are developing and their employees who have developed expertise in areas of interest. For example, Google acquired several robotics startups that had developed what was considered to be as the best intellectual property in this sector. Similarly, GM acquired Cruise Automation and Uber acquired Otto for the autonomous vehicle expertise the teams of these two startups had developed.

e. **Partner**: The Innovation Outpost can collaborate with startups in order to develop a disruptive new solution using the start-up's innovations along with those developed by corporation. In other words, partnering enables the *co-innovation efforts* between the corporate parent and startups. For example, a few years ago Mercedes had partnered with Tesla to develop batteries for electric vehicles.

The Innovation Outpost:

1. **Focuses** on the strategic innovation issues selected by the corporation. For example, technology and business model innovations related to autonomous vehicles and on-demand Mobility Services.

2. **Decides** which startups to invest in, partner with, or acquire in order to address the selected innovation issues and **selects** which internally and externally generated ideas to incubate and identify the teams to incubate them with. The selected startups and teams to be incubated represent *low-cost H3 experiments* for addressing the established corporate innovation goals. The Innovation Outpost **maintains** and **supports** these efforts outside the established corporate structures, policies, and culture, protecting them from the

corporation's "antibodies" that have a natural tendency to want to eliminate them. In this way, the more promising initiatives have a higher probability to succeed.

3. **Manages** the startup H3 portfolio(s) it creates, **selects** which of these startups to continue supporting, **grows** them into emerging H2 businesses, **staffs** these businesses, and **positions** them in the market. Again, by keeping such nascent businesses outside the established corporate structures, it protects them and offers the environment where they can continue to grow into the next-generation corporate business units. For example, consider Ford's new Smart Mobility business, and GM's Maven business.

4. **Collaborates** with corporate management to establish the criteria for launching the more successful of these emerging H2 businesses as full-fledged H1 next-generation corporate business units, as well as the criteria under which such units will remain independent or "re-enter" the established corporate structure.

Deciding which new efforts/experiments to include in its portfolio and how to manage the resulting portfolio is an *ongoing process* performed by the Innovation Outpost. The Innovation Outpost is the organization where the automotive industry incumbents can start working with start-ups to experiment with big data and machine intelligence applications and test business models for their monetization.

A fully developed Innovation Outpost consists of five groups (shown in Figure 10): Innovation Outpost (IO) R&D, Corporate Venture Capital (CVC), Startup Incubator, Business Development, Startup Corporate Development.

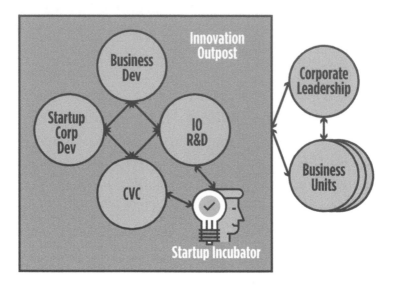

FIGURE 10: THE STRUCTURE OF THE INNOVATION OUTPOST. FIGURE COURTESY OF EVANGELOS SIMOUDIS

The Innovation Outpost has a single leader who reports directly to the company's CEO. The Outpost's leader must:

- Empower and ensure that the groups comprising the Outpost collaborate effectively and continuously, blending their efforts rather than operate as silo organizations.
- Ensure that the Outpost interacts constantly with the corporate R&D organization and the various business units in order to keep them updated on the Outpost's portfolio and confirm that the efforts remain relevant to the strategic innovation goals the Outpost is tasked to address.
- Have the constant support of the corporation's executive management and even its board of directors.

It is very instructive to understand how 3-Horizon Corporations such as Google, Facebook, and Amazon but also corporations such as Qualcomm and GE and even earlier stage companies like Uber have incorporated these principles into their innovation work. For example, Google Ventures, one of the most active corporate venture investors, closely collaborates with the company's corporate and business development organizations, as acquisitions of companies like Nest clearly demonstrate. Qualcomm has used its Innovation Outpost to address one of its strategic opportunity areas: robotics. The company's CTO led the innovation initiative (single leader). The initiative engaged four internal groups, including R&D and corporate ventures. The focus on robotics developed when the venture group collaborated with R&D to find adjacent markets for Qualcomm's Snapdragon processors. The corporation focused on robotics because that field can use many of the technologies Qualcomm has already developed for mobile phones, such as motion sensors, GPS, GPU, and computer vision. Qualcomm established short-, medium-, and long-term goals for this innovation initiative, with the associated timelines, and then addressed them strategically. In addition to starting a new R&D initiative,[122] Qualcomm acquired KMel Robotics[123] to address its short-term market opportunity. It invested in 3D Robotics[124] to address its medium-term goal (ROI in 4–6 years). To address its long-term innovation goal (ROI in 7–10 years), Qualcomm established the Qualcomm Robotics Accelerator[125] in collaboration with Techstars,[126] where it is currently incubating 10 startup teams. Finally, Uber has followed a similar blending approach with its autonomous vehicle efforts. In addition to continuing its own internal research, it acquired assets from Microsoft and also the startup Otto.

6.3 INHIBITORS TO APPLYING THE STARTUP-DRIVEN INNOVATION STRATEGY

We have analyzed the innovation efforts in several large corporations. We have identified several reasons that inhibit corporations from either launching H3 innovation initiatives at all, or from launching enough experiments to be in the position to disrupt and guard against disruption. These include:

1. Executive management's "short-termism." Corporate boards, CEOs, and senior executive management teams are engaged in well-documented *horizon conflict*[27] that places the need to always provide shareholder value front and center. The holders of big blocks of a corporation's shares, e.g., hedge funds, mutual funds, want value in the short term and influence corporate behavior, now called **short-termism**, that lead to such value, e.g., stock buybacks. This is causing corporations to remain locked in their existing products and business models and fail to see potential disruptions in time. To alleviate the short-term pressure for increasing shareholder value, corporations have been engaging in a buyback binge and large-scale acquisitions (Figure 11) that aim at efficiency improvements rather than invest in innovations that have the potential of providing value in the long term.

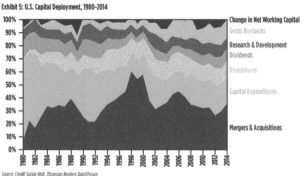

FIGURE 11: U.S. CORPORATE CAPITAL DEPLOYMENT 1980-2014. IMAGE COURTESY OF CREDIT SUISSE HOLT, THOMPSON REUTERS, DATASTREAM.

2. Middle-management resistance. Typically the senior management aspires to adopt particular innovations created by startups as well as the innovation methods developed and used by startups and their ecosystem in the process of creating these innovations. Additionally, the rank-and-file employees on a daily basis encounter a variety of problems that need innovative solutions and may be eagerly looking to apply novel approaches to address them. Middle management, on the other hand, either because they resist change in general or see innovation as a threat, largely resist them.

3. Corporate management's realization that technological and other types of H3 innovations take longer to adopt, develop, and roll out to market than innovators typically believe. This can be due to either corporate organizational inertia or market inertia.

4. Incumbents' beliefs. Incumbents in each industry believe that their established business models will continue to work well into the future and fail to foresee the impact of emerging technologies and business models. The disruption of the music industry from digital streaming media and the disruption of brick-and-mortar retailing through ecommerce are well documented. Typically corporations look into disruptive models once they start feeling the effects of disruption and, by then, it is too late. The automotive industry's potential disruption from new technology (ACE vehicles) and new business models (on-demand Mobility Services) is a case in point.

5. Labor and special interests in labor-intensive and regulated industries. The disruptions caused by startup innovations can have profound impact on labor-intensive industries such as automotive, manufacturing, logistics, agriculture, etc. As noted by the World Economic

Forum,[128] the technological and business model innovations that are impacting many industries, including the automotive industry, e.g., robotics, big data and machine intelligence, digitalization, etc., will have profound impact on the future of jobs. For example, consider the reaction of car dealers to Tesla's direct-to-consumer sales model.[129] In 2015 the automotive industry employed 1.5M people and impacted the employment of 7.25M people, including 1.1M people[130] who are employed by car dealers round the country.[131] Therefore, it should not be surprising that car dealers strongly resist the broader adoption of Tesla's direct-to-consumer sales model.

7.0
AUTOMAKERS MUST RETHINK THE WAY THEY INNOVATE

For more than fifty years, Silicon Valley has been at the forefront of many technological and business model innovations created by the startups that call the area home. Today, Silicon Valley finds itself again at the forefront, this time because of startups working on ACE vehicles, technologies such as big data and machine intelligence, on-demand Mobility Services, and business models such as ridesharing that are disrupting the automotive industry. Only this time Silicon Valley is not alone. It is also joined by startups in Israel, China, and other innovation clusters from around the world.

But while we may acknowledge that startups are becoming the driving force behind innovations that have the potential to disrupt or at least significantly transform the automotive industry, the industry's incumbents are not standing still. First, they have continued to increase their R&D budgets. Second, in the last couple of years more automotive industry incumbents established Innovation Outposts (IOs) in Silicon Valley and other innovation clusters, as well as expanded the operations of research centers they had set up a few years ago in Silicon Valley. Though they represent a positive step, these efforts are not yet commensurate to the potential opportunity and disruption risk.

7.1 INTERNAL R&D ALONE CANNOT CREATE THE NEEDED INNOVATIONS

During the last four years, automakers have been increasing their internal R&D investments. Today the automotive industry invests about $110B annually[132] in the areas it classifies as R&D, that were shown in Figure 2 of Chapter 2. Figure 12 shows the top 20 R&D spenders in 2015, based on data compiled by PwC,[133] where we see that five of the top 20 companies are incumbent automotive OEMs.

The majority of the automotive industry's R&D investments are made to *protect existing business models,* i.e., protect their H1 businesses, by focusing on:

1. Sustaining innovations (e.g., improving vehicle performance, improving driver ergonomics but not radically re-thinking the driver experience; compare the Google Maps experience vs. the in-dash navigation systems of 2016 model year passenger cars).

2. Innovations necessary to comply with government regulations (e.g., increasing the use of plastic and aluminum components to make cars lighter and able to utilize smaller engines thus decreasing their CO_2 emissions). Admittedly, some of the innovations necessary for compliance with government regulations can be disruptive, e.g., Ford's cost-effective use of aluminum in conjunction with new engine technology to make its F150 truck lighter and less polluting.

Rank	Company	2015 R&D Spend (US$ billion)
1	Volkswagen	$15.3
2	Samsung	$14.1
3	Intel	$11.5
4	Microsoft	$11.4
5	Roche	$10.8
6	Google	$9.8
7	Amazon	$9.3
8	Toyota	$9.2
9	Novartis	$9.1
10	J&J	$8.5
11	Pfizer	$8.4
12	Daimler	$7.6
13	GM	$7.4
14	Merck	$7.2
15	Ford	$6.9
16	Sanofi	$6.4
17	Cisco	$6.3
18	Apple	$6.0
19	GSK	$5.7
20	AstraZeneca	$5.6

FIGURE 12: TOP 20 CORPORATE R&D SPENDERS IN 2015

With its high barriers to entry because of the capital required, the automotive industry for many years had developed a false sense of security that drove its R&D investment strategy. Investments in technologies, such as big data and machine intelligence, that can lead to over-the-horizon (H3) innovations, such as ACE vehicles, were not top R&D priority.

Because of the way the industry has been allocating its internal R&D budget and the types of innovations it has been introducing to the market, broadly it is not viewed as innovative. Figure 13 shows the results of a survey, also conducted by PwC,[134] where executives from a variety of industries were asked to identify the top 10 most innovative companies of 2015.

Rank by Innovation	Company	2015 R&D Spend (US$ billion)
1	Apple	6.0
2	Google	9.8
3	Tesla	0.5
4	Samsung	14.1
5	Amazon	9.3
6	3M	1.8
7	GE	4.2
8	Microsoft	11.4
9	IBM	5.4
10	Toyota	9.2

FIGURE 13: PWC SURVEY RESULTS OF THE TOP 10 MOST INNOVATIVE COMPANIES IN 2015

Notice that Tesla Motors, one of our four disruptors, and Toyota (the only incumbent) are the only automotive companies included in the ranking. This is actually a slight improvement from the 2014 survey,[135] which included only Tesla in the list of the top 10 most innovative companies. Moreover, none of the Tier 1 parts suppliers, corporations like Delphi, Bosch, and Denso, are included in this list.

Figure 14 shows the results of a similar 2015 innovation-leader survey conducted by BCG.[136]

Rank by Innovation	Company
1	Apple
2	Google
3	Tesla
4	Microsoft
5	Samsung
6	Toyota
7	BMW
8	Gilead
9	Amazon
10	Daimler

FIGURE 14: BCG SURVEY RESULTS OF THE TOP 10 MOST INNOVATIVE COMPANIES IN 2015

The results of this survey for 2015 are also a little better for automotive incumbents than the corresponding 2014 survey.[137] While the automotive company included in the 2014 survey was Tesla, the 2015 survey, in addition to Telsa, includes Toyota, BMW, and Daimler. This list also does not include any of the Tier 1 parts suppliers.

Our analysis of the automotive industry's approach to R&D, combined with the results of the PwC and BCG surveys lead us to conclude:

1. High R&D investments do not eliminate the disruption risk from ACE vehicles and Mobility Services, if they are not focused on the key technologies that enable these disruptive innovations. In Chapters 3 and 4 we discussed the technologies that enable ACE vehicles and on-demand Mobility Services, focusing particularly on big data and machine intelligence. The incumbents' R&D investments in these technologies have been low, compared to the magnitude of the opportunity and the fallout from the potential disruption. For context, consider that GM stopped producing its electric vehicle EV1 in 1999 (a longer analysis is provided here). With their recent investments in innovation clusters such as Silicon Valley (discussed in the next section) automotive industry incumbents are trying to catch up.

2. The allocations of the R&D budget are a reflection of the automotive industry's process-oriented culture and highly hierarchical structure. Culturally, incumbent automakers prefer to, at best, be *fast followers*, rather than *first movers*, such as Tesla, Google, Apple, and Amazon. They perceive that in this way they are reducing the risk associated with early technology adoption. The traditional notion of fast follower has worked well in times of normal market conditions when paired with the automotive industry's practices that are defined by long development cycles. However, in the presence of *accelerating innovation in technologies that can disrupt the automotive industry's existing business models*, the notion of fast follower must also change. Fast follower no longer means that the corporation can afford to wait until a technology, or business model, is accepted by the early

adopters, implying a few years after the innovation's introduction. Instead, a fast follower may need to make adoption decisions within months and take on more risk. Failure to do so may imply losing significant market share that could become expensive, or at times impossible, to regain. Understanding how to enter a more speculative market opportunity, operating under the uncertainty such a move entails and evaluating the risk associated at various stages of this process represent the type of know-how that automotive industry incumbents can learn from working in startup ecosystems and using innovation strategies such as the Startup-Driven Innovation Strategy.

7.2 ESTABLISHING A PRESENCE IN A STARTUP ECOSYSTEM ALONE DOESN'T IMPLY RETHINKING INNOVATION

In the same way that high R&D investments alone do not make a corporation innovative, just establishing a presence in a startup ecosystem does not imply that the corporation has changed its innovation model and is ready to take on the risks associated with over-the-horizon innovations. Even before they started realizing the potential of their industry's disruption due to ACE vehicles and Mobility Services, some automakers had established operations in Silicon Valley. In the last couple of years, this trend has accelerated, and today several automotive OEMs and suppliers have also established Innovation Outposts in Silicon Valley and more recently in Israel.

Figure 15 shows the automotive industry companies that have established Innovation Outposts in Silicon Valley and also depicts the functions performed by each outpost, as these were described in Chapter 6 and shown in Figure 10. Today these corporations employ about 600 people in Silicon Valley.

Corporate Venture Capital	IO R&D	Incubator	Business Dev
BMW	BMW	BMW	BMW
GM	GM	Ford	Johnson Controls
Volvo	Daimler	VW	Faurecia
Nissan (via WiL)	Ford	Chrysler	PSA
Delphi	VW	Bosch	
Bosch	Delphi		
Nokia (Connected Car)	Bosch		
Hyundai	Honda		
SAIC	Nissan		
	Toyota		
	Volkswagen		
	Valeo		
	Mazda		
	Alpine Electronics		
	Denso		
	Continental		
	Ericsson (Connected Car)		

FIGURE 15: SILICON VALLEY-BASED INNOVATION OUTPOSTS ESTABLISHED BY THE AUTOMOTIVE INDUSTRY

It is easy to note that the majority of the corporations have only established R&D labs in their Silicon Valley Innovation Outposts. Figure 16 shows the areas where these Innovation Outpost R&D labs are focusing.

Company	IO R&D Areas of Focus
BMW	• Connected car • Autonomous car • Materials • UX
Mercedes	• Autonomous car • Electric drive • UX
VW	• Connected car • UX
Ford	• Connected car • Autonomous car • Big data • UX • Multimodal transportation
GM	• UX • Autonomous car
Nissan/Renault	• Connected car • Autonomous car • UX • Big Data
Toyota	• Connected car • Electric car infrastructure • Big data
Honda	• Connected car • Big data • UX • Security
Denso	• Autonomous car • Big data • Security
Continental/Elektrobit	• Connected car • Autonomous car • Mobility services
Delphi	• Connected car • Autonomous car
Alpine Electronics	• Connected car • Cloud computing
Bosch	• Connected car • Autonomous car • Big data • UX • Energy storage
Mazda	• Autonomous car • UX
Valeo	• Autonomous car • UX

FIGURE 16: AREAS OF FOCUS IN AUTOMOTIVE INDUSTRY'S SILICON VALLEY-BASED R&D LABS

Based on number of employees working in these Innovation Outposts and the data in Figures 15 and 16, we can draw the following conclusions:

1. The good news one can conclude from this data is that several of the automotive incumbents' Innovation Outposts focus on autonomous vehicle, connected car, and user experience (UX) technologies. The not so good news is that only few of the Outposts (Ford, Toyota, Renault, Denso, and Bosch) are working on big data and machine intelligence.

2. Fewer corporations than I had expected have established venture investment, startup incubation, and business development functions in their Innovation Outposts so that they can try to understand business model innovation and the best practices of these startup ecosystems. In the Innovation Outposts that include more than one of these operating functions, the efforts are often uncoordinated and are leading to interactions that can confuse startups and their venture investors on how to best work with the corporation. By contrast, in other industries (e.g., telco, consumer electronics), more corporations have established larger and more complete Innovation Outposts (e.g., Verizon, Telefonica, Orange, Samsung). BMW, Mercedes, Ford and GM are starting to work with startups in a promising way particularly in Silicon Valley, and their Innovation Outposts appear to be collaborating with corporate R&D, as well as other organizations.

3. Based on the small number of people they employ, most of these Innovation Outposts can only be involved in technology scouting and evaluation. Few are engaged in new technology development. At their current level, and over the long term, these efforts will prove

insufficient given the magnitude of the potential disruption from startup innovations in ACE vehicles and on-demand mobility. There are several reasons for this conclusion:

a. Effortless startup creation since we have "cracked the code" on how to create startups with small initial capital requirements;

b. Amply available and low-cost funding options from angels, institutional venture investors, and family offices that are investing aggressively in startups working in these areas as they sense the opportunity for outsized returns driven by the potential of the anticipated disruption and early exits of startups like Cruise Automation, Otto and others;

c. The impact of the changing model from one that relies exclusively on car ownership to a hybrid one that blends car ownership with car access;

d. Software- and big data-centric products that lead to differentiated user experiences and shorter timelines to market than the automotive industry's product development cycles.

The steps the incumbents are starting to take must be accompanied by more fundamental changes on how they approach innovation and the ongoing role of startups, particularly in identifying, testing, and deploying relevant over-the-horizon innovations.

Ultimately, the success of their efforts hinges on automakers becoming ambidextrous organizations. They need to continue executing well on their existing business models, while at the same time adopting a culture that embraces continuous innovation that combines their internal efforts with open innovation, often driven by startups.

4. The staffing of the Innovation Outposts will need to significantly expand and its composition re-considered. As we have observed the progress of Innovation Outposts established by corporations in other industries, the success of these organizations both on a stand-alone basis but also with regard to their ability to effect change in the corporate parent depends on the employees who staff them. The Innovation Outpost's employees must not be a replica of those employed by the corporation's business units. Moreover, there must be a genuine attempt to organize Innovation Outpost's employees using a less hierarchical and more agile organizational structure which could ultimately become the blueprint for the next-generation businesses (H2 and H1) that can be launched from the IO's efforts. The automotive incumbents' Innovation Outposts in Silicon Valley today employ a small number of primarily technologists. It is unclear yet whether these groups in their present size, structure, and composition could form a critical mass and impact their corporate parents' thinking regarding the actions that must be taken in order to avoid being disrupted and be in a position to disrupt. As they stand today, these groups are just too small and not diverse enough to have a transformational impact on their parent corporations in light of this disruption. Finally, the executives leading these Outposts still are not senior enough in the organizational hierarchy of corporate parent. As a result, they are not able to take the bold, strategic actions that are often necessary as they try to compete with the rapidly evolving area of ACE vehicles and on-demand Mobility Services.

In order for their corporate parents to capitalize on the opportunity presented by next-generation mobility and the driverless future, the automotive incumbents' R&D efforts must be re-thought become better coordinated with the efforts of their Innovation Outposts that must start to strategically focus on the application of big data and machine intelligence to ACE vehicles and Mobility Services. Incumbents must dedicate meaningful financial and human resources on this endeavor as they consider the variety of innovation types that are possible. Data from recently conducted surveys shows that consumers would trust driverless vehicles that OEMs develop in collaboration with technology companies, including startups, more than the vehicles either group develop on its own. For this reason, automotive incumbent efforts on next-generation mobility must blend in-house development of over-the-horizon technologies with co-innovation, partnerships, acquisitions, investments, and incubation of startups.

8.0
AUTOMAKERS MUST MAKE
BIG DATA A STRATEGIC IMPERATIVE

Many analogies have been used to describe the ACE vehicle: robo-car, four-wheel smart device, etc. While these may be appropriate descriptions for the general public, as we mentioned in Chapter 4, the autonomous connected vehicle should also be considered an insightful application that interfaces with a variety of hardware, e.g., sensors, actuators to interact with its environment.

While offering ACE vehicles may eventually become table stakes, by designating big data a strategic imperative now and developing a core competence around its exploitation using machine intelligence, automakers have a unique opportunity to play a disrupting role on an ongoing basis. Through big data they will be able to provide strong and superior customer experience as the next-generation mobility model evolves. The innovations generated from the application of machine intelligence on transportation-related big data will accelerate the disruption of the value chain, necessitate new regulations, and make possible the emergence of new leaders. Incumbent automakers must develop such expertise regardless of the level of driving automation they offer, or plan to offer, in their vehicles. But they are not starting from a great position.

8.1 DATA EFFORTS IN INCUMBENT AUTOMAKERS TODAY

Leaving aside any big data that may be produced during the vehicle design and manufacturing processes and used for the analysis of the supply chain and to understand quality issues, the only other data that incumbent automakers are familiar with is the data generated by the systems embedded in their vehicles and is used to understand a vehicle's health.

Over the years, an increasing number of Engine Control Units (ECUs) and other microprocessor-controlled subsystems[138] have been embedded in cars. However, there are three important points to note about these subsystems:

1. They are provided by various suppliers, e.g., Delphi, Denso, Magna, Bosch, Continental, etc., rather than the OEMs themselves. Moreover, they are integrated into the vehicle on an ad hoc basis, i.e., without a well-defined system architecture that can be updated appropriately as more microprocessor-controlled subsystems are added to new vehicle designs. We are now to the point where certain vehicles may include more than 100 microprocessors on-board. Interestingly, because of this approach to vehicle design and innovation, the Tier 1 and Tier 2 suppliers have better appreciation and knowledge of the data generated by the vehicle than the OEMs themselves.

2. Only a small percentage of the data vehicles generate is accessible, typically via the ODB interface,[139] and utilized. Most often the accessible data is used to provide variants of a vehicle's health report. It ranges from the familiar "check engine" light seen in low-end vehicles, to the graphical reports provided in the infotainment system of higher-end vehicles, or more recently on mobile phones.[140] As a result, to date third parties have not been able to develop interesting applications, though many startups are trying, e.g., Automatic, Zubie.

3. While these systems generate data, they don't generate data with the same variety, velocity, and volume as next-generation vehicles and the infrastructure they will operate in, i.e., *they don't yet generate big data*. As a result, incumbent automotive OEMs and most of their suppliers don't yet realize the strategic importance of big data.[141]

Because most of the vehicles on the road today don't generate big data, automakers are not used to collecting big data. And collecting big data is just the beginning. Exploiting it effectively is significantly harder. Making big data into a core competency won't come naturally and won't come easy because automakers must appreciate:

1. The difference between customer-centric on-demand mobility, where data exploitation is at the center, and car manufacturing, where data exploitation may be useful but is not central. With ACE vehicles and on-demand Mobility Services automakers will need to start considering data that is necessary for the autonomous vehicle's navigation, as well as data relating to the overall customer and user experience. Automakers traditionally have not captured these types of data. Furthermore, because their primary focus is on the driver, customer and user experience is typically synonymous with

driver experience. With the possible exception of GM's OnStar service, only recently have other automakers, such as Ford,[142] started to collect and experiment with these types of big data.

2. The difference between ACE vehicles, where data exploitation makes navigation and many other functions possible, and the cars they've been designing and manufacturing for the past 100 years.

3. That such competency cannot be developed overnight. It requires a long-term approach. In some areas such as autonomous vehicle navigation, it may already be too late for incumbent automakers to be able to disrupt broadly with big data and machine intelligence because they lack the necessary data assets, e.g., mapping data. Mercedes, BMW, and Audi may be able to be fast followers now that they have acquired HERE,[143] a company that is currently in the process of producing high-definition maps. However, even they will need to improve their ability to exploit this type of big data. Ford through an investment in CivilMaps,[144] a startup developing a new high-definition mapping platform, will be testing whether it can also become a fast follower in this area.

4. That their efforts to develop competence in big data with the associated critical mass of employees, regardless of whether they are led by corporate R&D, the Innovation Outposts, or other organizations, though encouraging are not sufficient. GM's experience with OnStar[145] may represent an exception. GM established the OnStar service in 1995,[146] and since that time it has been collecting data from the service's subscribers and the vehicles they operate. While the OnStar data being collected

has not been exploited to its fullest, nonetheless it is providing GM with an advantage over its competitors. In addition to GM, a few automakers have been taking initial steps to gain general big data exploitation expertise, primarily through acquisitions of, selective investments in,[147] and partnerships[148] with relevant companies, and by starting to hire data scientists[149] and setting up big data infrastructures[150] internally. Automotive suppliers are also working aggressively to develop expertise in big data exploitation. A perfect example of this point is NVIDIA, which aspires to become one of the leaders in data processing and machine learning[151] for self-driving cars with Level 4 autonomy. It is unlikely that NVIDIA will share its expertise with its automaker partners. At best, these partners will access this expertise through the NVIDIA products they license.

To overcome these limitations and translate their big data access into a true disruptive competence and long-lasting asymmetric advantage that will place them in a leadership position to deliver next-generation personal mobility solutions, requires automotive incumbents to designate big data as a strategic imperative. Under such an imperative each incumbent must develop the necessary **strategy**, **technology**, the right **organization** and an innovation-driven **culture**. In particular, automakers must:

1. **Develop and own** next-generation mobility's big data strategy.
2. **Create** a unified architecture that combines computing with big data and machine intelligence.
3. **Establish** and **enforce** data-ownership rights among the appropriate constituencies.

4. **Create** a data-sharing culture and **establish** a broad and dynamic partner ecosystem with other experts in big data and machine intelligence.

These activities can be led by each automaker's Innovation Outpost but ultimately, and in order to be successful, they must involve the entire corporate leadership.

8.2 OWN THE BIG DATA STRATEGY BUT NOT ALL THE DATA

The acknowledged disruptors of next-generation mobility, e.g., Google, Tesla, Uber, but also startups like Faraday Future, Atieva, Renovo Motors, and Divergent3D, have realized the importance of big data and its exploitation through machine intelligence, and have created detailed data strategies from the moment they started to design their vehicles, the operating systems that will power next-generation vehicles, and/or their mobility services. They have also established sizable data science organizations that are working closely with the product teams for the refinement and implementation of these strategies.

Incumbent automakers must develop their own data strategies that specify:

1. What data to collect to enable autonomous driving;

2. What vehicle and driving environment data to combine with external data, e.g., mapping data;

3. How to apply machine intelligence to analyze this data, and how frequently to update the predictive models being employed;

4. What data to capture in order to understand driver and passenger preferences and vehicle performance;

5. What data will give them 360-degree understanding of the vehicle, its internal and surrounding environment, and its occupants;

6. Which data-driven applications their strategy must include, which will be developed internally, which externally, and which will be acquired, which will be custom, which will be off the shelf;

7. How all these applications will be interfacing and interacting with third-party applications.

Automakers must not only develop big data strategies that address all these areas, but they must own them and drive their execution to their suppliers, partners, and customers.

Owning and driving the execution of the big data strategy will require incumbent automakers to control the parts of their supply chain that deal with data-generating components. When this is not possible they will need to establish agreements with their suppliers and partners that give them broader access to data generated in their vehicles. Today they take a different approach to their supply chain. Over the past few years, automakers became systems integrators, and in some instances they have even been acting more like contract manufacturers; think Foxconn or LG, both of which are already working with the automotive industry. For each new vehicle they design to address perceived market needs and adhere to government regulations, e.g., safety, emissions, they issue to the supplier network a master specification. With the exception of internal combustion engines that most OEMs design and manufacture, the suppliers provide the components and subsystems that conform with the issued specification. OEMs assemble the procured components into a vehicle through a highly optimized and efficient process. With the move to electrified vehicles even engine design and manufacturing may be supplied by third-parties unless incumbents elect to develop the necessary expertise.

The approach used by OEMs today means that much of the innovation and knowledge on how to address these requirements go to the supplier, rather than the automaker. Similarly the data produced by each such subsystem, e.g., infotainment, is owned , and thus exploited, by the supplier, rather than the automaker. For example, consider the value Tesla derives from the data it collects by having full control over the proprietary infotainment system installed in its vehicles and the ways it can exploit this data to provide a personalized driver and passenger experience versus what any other automaker that incorporates an infotainment system from a supplier such as Bosch can provide.

In designing their big data strategies, automakers should strive to collaborate with their Tier 1 suppliers, some of whom have embarked on creating such strategies, and with leading automotive startups, but also consider the big data strategies of leading enterprises in other industries such as telco, aerospace, manufacturing, logistics, and even financial services, insurance, and consumer packaged goods.

8.3 CREATE A UNIFIED ARCHITECTURE COMBINING COMPUTING AND BIG DATA SERVICES

ACE vehicles will have significantly more computing power running software that will be orders of magnitude more complex than the software in today's vehicles. These vehicles will be equipped with thousands of sensors to detect everything about road (infrastructure conditions, traffic conditions), passenger, vehicle (from engine, to actuators, controllers, tires, cabin, etc.) and environment conditions. This prospect will necessitate that automakers create for the first time a modern, adaptable architecture that will integrate from the ground up the vehicle's data and computation requirements including those of the machine intelligence applications running on the vehicle.

At the heart of such architecture must be a new operating system. An example of where we may be heading is the TeslaOS, the operating system running today on the Tesla vehicles. The operating system running on tomorrow's ACE vehicle must be **open, extensible** through well-described and well-behaving APIs, and **data-centric**, allowing the flow, utilization, and exploitation of big data through well-specified and extensible software buses. In order to comprehensively address all the data types needed by the various intelligent applications, this architecture must also utilize the data framework presented in Chapter 4.

To develop know-how about such architectures, incumbent automakers should establish long-term strategic partnerships with startups like Renovo, Harbrick, and OSVehicle that bring fresh visions of vehicle operating systems and have already started working in this area with next-generation automakers and their suppliers.

A unified architecture must include services for:

1. Collecting data. To be effective, data must be collected regardless of whether it comes from a subsystem, a sensor, the passengers, the transportation infrastructure, other vehicles, or the environment and whether it relates to the operation of the vehicle or any other aspect of the transportation experience. Capturing the *rules of data pertinence, data storage location,* and *data half-life* as outlined in the framework presented in Chapter 4 will be an important piece of intelligence these architectures should possess.

2. Managing the collected data. Automakers have developed sophisticated data centers that only support their internal operations.
In-vehicle data management is being used today only for the data
captured from the small number of the vehicle's embedded systems
being monitored, the mapping data that is being used by the vehicle's GPS component, passenger contacts data, and a certain amount
of entertainment content.

ACE vehicles have more complex data management requirements.
They will need to manage the torrent of data generated by the vehicle
itself, the infrastructure, other vehicles, a variety of external sources
and communicated to the vehicle to be used for its navigation needs,
as well as passenger infotainment and other passenger-related data.
Periodically, data with longer half-life, will need to be moved from
the ACE vehicle to one or more of the clouds being used. We envision clouds owned by OEMs, by value chain partners, such as Tier
1 suppliers and mobility application providers, e.g., Transportation
Network Companies, or third-party cloud infrastructure providers
such as telcos, e.g., Verizon, IT vendors, e.g., IBM, Microsoft, or an
internet-related provider, e.g., Amazon's AWS.

3. Fusing the collected data. Because of the quantity and variety of
low-level data collected by the various sensors of ACE vehicles,
it will be necessary to perform extensive data fusion onboard the
vehicle.[152] Performing data fusion will require specialized software
to make the resulting information very approachable and digest-
ible to passengers, particularly in vehicles with up to Level 4 driving automation since these require, or at least accommodate, the
driver. The driver or passengers of such vehicles should not need

the equivalent of pilot training before being able to fully take advantage of the opportunities afforded by the vehicles.

4. Communicating data. Discussions about connected cars typically invoke examples relating to entertainment and productivity applications, e.g., passengers accessing streaming music, video, email, etc. While such applications have data communication needs, ACE vehicles, in addition to this type of data, must be able to communicate big data about the vehicle's inside and surrounding environment, its operating status, as well as deliver operating system and application software updates, and machine intelligence-based predictive model updates.

As we have seen through the much simpler by comparison experience of accessing applications from our smartphones, the bandwidth needed to achieve acceptable performance anytime and anyplace is a function of the cellular networks used. However, even in this early stage of developing an deploying ACE vehicles we have started to realize the speed and bandwidth offered by such cellular networks, even when 4G technologies are being utilized, are inadequate for transmitting the volumes of data captured and needed by such vehicles, particularly as we envision a world with millions of autonomous connected vehicles.

Wireless carriers, like Verizon, are experimenting with hybrid data-transmission methods that utilize both cellular and Wi-Fi networks. The unified big data architecture must take into account such lessons and accommodate the solutions carriers are devising. For this reason, it will be important for automakers to establish strategic partnerships with wireless carriers globally. In addition the architecture must incorporate data communication policies that establish the rules under which each type of data is transmitted. For example, should vehicle status data move to the cloud when the vehicle is at rest at a home base or at other times as well?

5. Software and predictive model management. This is a very difficult issue and an area of active research and debate. In ACE vehicles, Over The Air (OTA) updates are of paramount importance because in addition to the various software components that will need to be frequently updated, the predictive models used by such vehicles must be frequently updated, too, as companies like Tesla and Google are already demonstrating.[153]

The predictive models being used for ACE vehicle navigation are particularly complex because they are mostly based on deep learning approaches. Machine intelligence applications and associated models must be kept current through automatic updates in order to guarantee the vehicle's operating integrity and the passenger safety. The applications and models must always reflect the important new cases that are being encountered while these vehicles are used. The unified big data architecture must ensure that each such update doesn't create undesirable interactions with the performance of the software and the models that are not updated.

Software and model validation and quality assurance are issues that must govern the entire update process so that the right software and predictive models are always in use.

6. Securing data. Much has already been written about the risk of hackers accessing the data of ACE vehicles and causing various types of problems, from accidents to terrorism to financial blackmail of individuals, corporations, and even governments. As a result of this

risk we are already seeing the first venture investments in startups that develop automotive security solutions,[154] as well as the first acquisitions of cyber security companies that focus on automobiles.

The public's concern about automotive data security has been listed as one of the reasons that could delay the adoption of ACE vehicles.[155] To safeguard the devices that are important for the autonomous vehicle's operation, the data, software, and predictive models being used by the vehicle, the unified big data architecture discussed here must provide strong security services such as firmware anti-tampering protection, network security, anti-malware, data encryption, etc. To this end, automakers have recently published a set of best cybersecurity practices[156] that must be incorporated in the architecture.

Finally, the architecture must provide services to protect against data loss and enable data recovery. Today, when something happens to my smartphone and I lose data as a result, I can go to my computer, download a backup, and I'm able to restart my work. The data I lose in such an experience depends on the frequency and extent of my backups. The implications and ramifications if something equivalent were to happen to an ACE vehicle are far more dramatic. For this reason, the unified big data architecture will need to incorporate services that ensure the safe operation of the ACE vehicle even when data is lost or corrupted in order to gain and maintain the passenger's trust.

8.4 ESTABLISH AND ENFORCE DATA OWNERSHIP RIGHTS BUT NOT DATA WALLS

Four constituencies can claim ownership to the big data being created by ACE vehicles coupled with on-demand Mobility Services:

1. Automakers

2. Parts suppliers

3. Services providers; from the traditional, e.g., financial services and insurance, to the next-generation, e.g., ridesharing and carsharing companies, telcos, and utilities.

4. Consumers

It is impossible to think that a single constituency will be able to collect and own all the data that is necessary and prove useful in various transportation scenarios, particularly now that automakers are entering the Mobility Services business. It will be important for the corporations that want to participate in this emerging ecosystem to decide what data and information they will need and a which of this data they must own.

In addition to companies like Google and Apple that are collecting data through their mapping and infotainment systems, ridesharing companies like Uber and Lyft, or carsharing companies like Zipcar, generate and collect big data that the other corporate constituencies find interesting. We are also seeing the emergence of a very rich startup ecosystem working on Mobility Services with companies like MileIQ and Automatic.

The true value of this big data will come from the ability to effectively mesh elements from the data collected by each constituency and exploit the resulting sets using machine intelligence. For this reason, each constituency has to accept that while it has rights to certain data, no constituency will ever own all the data. Each constituency's rights have to be established, recognized, and safeguarded. There is significant and ongoing debate in the automotive world regarding data ownership. Today while automakers, and less so their parts suppliers, are not yet fully aware of the competitive advantage that big data might be providing them, they do not appear eager to share the data they have collected and that which they will be collecting. We are seeing early signs that they are already establishing so-called "walled gardens." Service providers, including some on-demand Mobility Services providers, have started taking a similar approach. For the most part, consumers are not even aware of the data that is being collected during their mobility activities and that which is possible to be collected in the near future. The automotive industry must try to learn valuable lessons from the experience telcos had with smartphones. Telcos also tried to create walls around the big data they were collecting from consumers. However, the limited ways they established for partnering with the then-developing mobile Internet and smartphone application ecosystem caused consumer backlash which was followed by government regulation. The beneficiaries of the telcos' approach became Google, Facebook, Amazon, and Apple.

Not unlike what has happened with Internet services, the participants in the automotive value chain, including consumers (maybe even starting with the consumers), need to know their rights with regards to each piece of data being collected, who owns each piece of data, under what conditions, what value its use will provide, and to whom, each piece of data can be used, and under what conditions its ownership can change. For example, if I as a consumer change cars and move to a different brand, what part of the data that the automotive value chain collected about me will move with me? Automakers will also need to think about what happens to the data when companies with which they partner declare bankruptcy or get sold, and reflect the right policy in the agreements they sign with their other big data partners. Internet companies had to make corresponding adjustments with regard to the data they capture.

8.5 CREATE A DATA-SHARING CULTURE AND ESTABLISH A PARTNER ECOSYSTEM WITH EXPERTS IN BIG DATA AND MACHINE INTELLIGENCE

By extensively utilizing big data and machine intelligence and paying attention to detail, Tesla has changed the conversation on the type of personalized experience car owners (drivers and passengers) should expect from an automaker. In the process, it has created a data-sharing culture and is building strong loyalty with the owners of its cars who appear willing to support it through thick and thin. Tesla has taken a lesson from Apple, Google, Facebook, and Amazon, four companies that obsess about connecting pieces of data, using it to better understand their consumers and tailor their services to provide the right customer experience, and sharing data with partners in order to enhance this experience. The personalized experience that Tesla offers has allowed it to build a brand that delights its customers.

As they attempt to establish a leadership position by exploiting big data from ACE vehicles and on-demand Mobility Services, automakers need to develop a **data-sharing culture** and accelerate their partnerships with companies that have strong data collection and exploitation DNA. Apple's recently announced investment in the Chinese Transportation Network Company Didi Chuxing[157] is another example of the importance and necessity for partnerships in developing data-centric transportation solutions even among market-leading companies. As on-demand Mobility Services are starting to play an increasingly important role in transportation solutions, companies that offer such services become ideal big data partners to automakers. By partnering with them, automakers will have the opportunity to understand consumers in far greater detail than they do today, including the consumer attraction to the business models associated with Mobility Services.

Ridesharing, carsharing, parking services, and potentially goods delivery services companies (particularly those operating in India and China since they make up for the lower-quality transportation and payments infrastructures in these two countries), represent the best initial candidates for such partnerships because:

1. Ridesharing, carsharing, parking, and goods delivery have emerged as the biggest/most popular and more successful Mobility Services. Companies like Uber, Lyft, and Didi Chuxing are establishing a global footprint. The insights extracted from the data they collect reflect this.

2. The companies offering these services are big data and machine intelligence companies, collecting and exploiting consumer big data with the same attention and rigor as Apple, Google, Facebook, and Amazon. They have already collected impressive data sets due to the scale they have achieved. Transportation Network Companies are expected to be the earliest adopters of ACE vehicles because of the anticipated positive impact such vehicles will have on their business model. And as they use ACE vehicles they will collect additional types of big data, ultimately potentially becoming leaders in driverless vehicle know-how.

3. Their data that can be used by high-definition mapping platforms. Even if they have their own proprietary access to such data, as Mercedes, BMW, and Audi have after acquiring HERE, the big data

provided by such Mobility Services companies, particularly real-time updates, can only improve their own data sets and therefore their vehicles' autonomous navigation abilities.

4. The data collected by ridesharing companies can be used in the design of next-generation vehicles to improve the customer experience. Remember that while automakers principally think about the driver, ridesharing companies principally think about the passenger. Consider Nissan's work in designing the NY taxis.[158]

5. By analyzing how many miles/year a person rides in an automaker's vehicles (through personal vehicles, carsharing, and ridesharing), automakers can offer loyalty programs with specific benefits. For example, for every 5,000 ridesharing miles in a particular automaker's vehicles, the person receives free cellular data to be used in their personal vehicle. Such offers could lead more consumers to sign up for car data plans, like the Tesla owners are required to do.

6. The automaker and its TNC partner can tie together driver and passenger data, e.g., when I travel to San Francisco, I'm a driver, but when I move around San Francisco using a ridesharing services such as Uber, I'm a passenger. In this way, automaker and TNC gain a more complete profile of each individual. Having access to such profiles, the automaker can segment each market and understand better the preferences of each segment.

Automakers are starting to make large investments in ridesharing and carsharing companies. GM's investment in Lyft,[159] Toyota's investment in Uber,[160] VW's investment in Gett,[161] and BMW's investment in Scoop[162] are early indications that automakers are starting to appreciate the importance of ridesharing and carsharing companies. In addition to seeing these services as a new, potentially lucrative, channel for selling their vehicles, they are starting to realize the important role these services will play in the new transportation model that is starting to emerge (so they see these investments as a hedge). But very few recognize that these investments also provide a great opportunity to partner around big data.

But why would such Mobility Services companies, particularly TNCs want to partner with the automakers? Let's consider the example described in Section 4.1 from Uber's perspective. Like an airline does with its planes, Uber always wants to understand which of the vehicles in its service are more efficient and allow for higher utilization, while providing the highest satisfaction to the passengers. Under the big data partnership, the automaker would provide Uber with the performance data collected from each vehicle used during my rides. Uber would use this data as it attempts to improve the ridesharing operating costs and, as a result, could decide to use for its service more of the automaker's vehicles. This is exactly what GM and Lyft could do as they develop their partnership. Finally, the data collected by the automaker about its car owners, including data collected through the automakers' concierge services, can be valuable to the Mobility Services company as it tries to understand how such owners use its service so that it can provide them with better and higher-margin transportation solutions.

8.6 IMPLICATIONS OF MAKING BIG DATA A STRATEGIC IMPERATIVE

In making big data a strategic imperative, automakers must:

1. *Embrace big data at their core.* Under a hybrid car ownership/car access model, automakers may need to own a larger part of the new value chain. For example, in the same way that today they own

financial services in order to be able to offer loans to consumers, in the future they may also need to offer car insurance to drivers of ridesharing services that use their vehicles, or maintenance and charging stations for the electric vehicles used by TNCs. This will require that they transition from being exclusively manufacturing and distribution companies. As part of this transition they must become information technology and insight generation companies. Offering insurance will require automakers to develop new risk models by utilizing big data, including data from several of their partners, e.g., telcos. Realizing such opportunities will require the data-sharing culture we discussed previously and will necessitate an important adjustment by the automakers that today tend to regard that they own all vehicle-related and driver-related data.

2. *Experiment and show patience in the process of monetizing the data.* While capturing and exploiting big data will require the incumbent automotive industry to develop the right expertise and a different innovation strategy and culture, monetizing the results of this exploitation (collected data, generated insights, intelligent data-driven applications) will come with its own challenges. Even if incumbents are able to provide innovative data-driven products and services, this may not lead to immediate and direct monetization opportunities as many next-generation mobility newcomers are already discovering. For this reason the overall big data exploitation effort, technology and the potential business models, should be considered as a series of *H3 innovation initiatives* and be driven by the efforts of the Innovation Outpost.

Industry analysts[163] estimate that the value of this data is worth approximately $100/vehicle/year and McKinsey estimates[164] the overall revenue from data-driven products and services to be in the range of $450–750 billion by 2030, based on a number of use cases. However, it is still too early to make definitive statements about the monetary value that can be derived from the exploitation of this data.

Automakers must be willing to experiment with various business-to-business and business-to-consumer monetization models during this phase, by testing them in conjunction with applications such as those presented in Section 4.1. Consumers have shown the willingness to pay for some Mobility Services, e.g., ridesharing, usage-based insurance, concierge, preventive maintenance, but not for every type of service. And in many cases the margins for the services that are being monetized are very thin. Consistent with the tenets of the Startup-Driven Innovation Strategy, automakers should be ready to *quickly* eliminate candidate models that are not succeeding while continuing to pursue the ones that show promise. In this effort it may be instructive to look for lessons from the use of online advertising technology by the telco industry. Telcos are employing sophisticated big data exploitation approaches in conjunction with online advertising platforms as they try to identify viable business models to monetize their data. Incumbent automakers should also consider Tesla's efforts to continuously experiment using the over-the-air updates to determine which new features and transportation experiences consumers will be willing to adopt and pay for.

3. *Exploit all available data to provide a better total user experience.*
As we move from car ownership to a hybrid model that blends car
ownership with car access, the quality of the user experience offered
to the consumer inside and outside the vehicle should become a first
priority for automakers. This is different from the total driver experience automakers traditionally worked on to improve. We have
discussed some of the technologies that are making autonomous
and driverless vehicles a reality and the complexity of these technologies. For consumers to gain trust in such vehicles, the technologies must not only become intuitive, they must become invisible.
Consider the effort consumers spend just to understand the features
of the complex infotainment systems found in today's cars. Now
consider how much more complicated this task will become in a
self-driving, and eventually driverless, car. As we saw in Chapter 7,
the Innovation Outposts of several automakers are working on user
experience projects. However, to date these don't exploit big data. By
exploiting big data and utilizing a combination of in-vehicle processing with cloud-based processing enabled by always-on broadband
connectivity, the automakers, in collaboration with partners, should
try to create the compelling in-vehicle user experiences that will be
necessary for such vehicles to be broadly accepted. Driver and passengers in the case of autonomous cars, and the passengers in the
case of driverless cars must feel stress-free and trust their vehicle to
the point where they will not have to pay attention to the vehicle's
operation and instead focus on other tasks such as entertainment. If
Tesla has shown us anything, it is that a superior in-vehicle experience that blends hardware, software and big data technologies is as

important as a general automotive experience that includes buying or leasing, charging or servicing the vehicle. But beyond the in-vehicle experience, the total user experience must also include the Mobility Services that are becoming part of the car access model. A significant part of this total experience will be driven by the insights derived from the exploitation of big data.

4. *Understand that small-batch vehicle manufacturing is economically feasible.* Additive Manufacturing enabled by 3D printing technologies is starting to change many industries, the automotive among them. Today it is being used by automakers and their suppliers primarily for rapid prototyping and in the production of smaller, non-critical parts. However, it has the opportunity to play an even more transformative role in the future, leading to the creation of cheaper, lighter, and safer products that are tailored to specific environments, e.g., urban, industrial, agricultural, hazardous, etc., and Mobility Services such as ridesharing and goods delivery. Along with the materials being used by 3D printers (such as polymers and metal alloys), and the printers themselves, big data will be a critical ingredient. For example, the big data collected by ridesharing and carsharing companies, can be used together with 3D printing technologies, and small, robotized assembly plants that utilize simpler supply chains to make economically feasible the manufacturing of specialized vehicles, e.g., ridesharing vehicles, in small batches and closer to the places where they will ultimately be used. Function-specific vehicles designed and manufactured this way can then be purchased and used by the ridesharing and carsharing companies themselves, further impacting positively their pricing

models. Small-batch manufacturing will impact not only vehicle design and pricing but also urban planning and design. Startups like Divergent3D and Local Motors are already experimenting with the combination of big data and small-batch-manufacturing approaches.

In the emerging and rapidly evolving model of personal mobility, automakers must make big data and machine intelligence a strategic imperative.

In addition to developing their expertise in this area, develop the right strategies, architectures and organizations, they must understand the implications of their actions and inaction on their future. The next chapter discusses five recommendations for disrupting with big data in a driverless future.

9.0
FIVE RECOMMENDATIONS FOR DISRUPTING WITH BIG DATA IN OUR DRIVERLESS FUTURE

I n this book I have argued that:

- Big data combined with machine intelligence are key ingredients in the future of mobility, particularly of urban personal mobility, that will increasingly depend on *Autonomous Connected Electrified* (ACE) *vehicles* and *on-demand Mobility Services*. They are key regardless of the level of driving automation being used.

- With many of the technologies enabling ACE vehicles becoming table stakes, and with Mobility Services leaders emerging from a group of global, well-financed startups, incumbent automakers must immediately designate big data as a strategic imperative. Important and differentiated value can be created from the exploitation of big data, providing the automotive industry with one of the best opportunities to lead and disrupt in the next-generation mobility market.

By acknowledging the strategic value of big data, incumbent automakers are accepting that they are in the information and insight generation business in addition to being in the manufacturing and distribution business.

- Big data in next-generation mobility is a Horizon 3 effort requiring the re-thinking of how the incumbent automotive industry in general and automakers in particular innovate.

In this chapter I provide five recommendations the automotive industry should consider once designating big data a strategic imperative.

Recommendation 1: Augment the R&D-centric corporate innovation model that today focuses on H1 and H2 innovations, with one that seeks to create, adopt, and further develop over-the-horizon innovations, and embraces other types of innovation in addition to technology.

The technology-centric, corporate innovation model has run its course, and now it may be effective to develop technologies that principally support existing (H1) business models and their extensions (H2). Though necessary, annual increases of R&D budgets will not change this reality. Innovation leaders such as Apple, Google, Amazon, and Tesla, which are considered 3-Horizon Corporations, have stopped relying only on the R&D-centric innovation model and utilize effectively the global startup ecosystem through venture investments, acquisitions, partnerships, and startup incubation.

The overall accelerating pace of innovation, the variety of technological and other types of innovations that lead to incumbent disruption in many industries, and the contributions of startups from around the world in creating these innovations necessitate that automotive industry incumbents augment or radically transform the existing corporate innovation model with one where startups and their ecosystems play a more pivotal role. The Startup-Driven Innovation Strategy proposes an innovation model for automakers to consider, and become 3-Horizon Corporations.

Recommendation 2: Assign to Innovation Outposts the mission to make big data and machine intelligence corporate core competencies and enable them to focus on more than technology innovation.

Many automakers believe that they are embracing startup innovation just because they establish an Innovation Outpost. In many instances they also strongly believe that these Outposts must be simple extensions of their internal R&D efforts. This means that the Outposts must focus on the development of technologies the innovation cluster is known for, but also corporate R&D is already be familiar with, e.g., ADAS sensors. As a result, in the case of next-generation mobility many corporate Innovation Outposts are working on technologies that have already become table stakes, rather than technologies and models that will enable their corporate parents to disrupt, e.g., big data for small-batch manufacturing. Big data and machine intelligence technologies must be a focus area in every automotive incumbent's Innovation Outpost. The big data strategy and architecture provided in Chapter 8 must be developed and deployed by the Outpost rather than the corporate IT organization.

While it is necessary for the automotive industry's Innovation Outposts to include technology development capability in order to be in the position to co-innovate with local startups, it is as important for these Outposts to include groups for business development, venture investments, corporate development, and startup incubation so that they can be in the position to make build vs. buy vs. partner or invest decisions. The Outpost's groups must be focusing on identifying, experimenting with and selecting the best big data and machine intelligence over-the-horizon (H3) technology, business model, and other types of innovations that will enable the automaker to disrupt in next-generation mobility.

Over time, the most promising of the big data and machine intelligence H3 innovation efforts supported by the Innovation Outpost should be provided with the appropriate resources and become the new H2 mobility businesses. These are the units that will start scaling the business models that will monetize big data and the insights derived from such data. Once they achieve adequate scale, these units will give rise to the next-generation corporate business units (H1). The automaker must be willing and ready to set up these businesses as new standalone entities and encourage them to develop new and entrepreneurial corporate cultures, rather than adopt cultures that are replicas of the corporate culture found in existing H1 and H2 business units. For this reason, the Innovation Outpost must be staffed with entrepreneurially minded employees who can easily transition to the new businesses, as appropriate. Automakers such as Ford, GM, Volvo, Daimler, and BMW[165] have already started establishing direct-to-consumer units to better focus on their next-generation mobility efforts, though it is not clear how they are shaping the corporate cultures of these units. These units can either serve as templates for other independent businesses the automakers can establish to focus on the monetization of products and services derived from big data and its exploitation, or their mission can expand enabling them to eventually also offer such products and services.

Recommendation 3: To derive maximum value from big data innovation the automakers must ensure the continuous **value exchange** and
collaboration between startups, the Innovation Outpost, and the rest
of the corporation, with the constant engagement of the senior executive management.

In addition to establishing one or more Innovation Outposts in
selected innovation clusters, with the appropriate set of operating functions and strategic areas of focus, including big data and machine intelligence, automakers must ensure that the Outpost's activities are continuously providing value to the startups and to the rest of the corporation.

Good coordination starts with the right reporting relations. For example, we have found that some automaker Innovation Outposts report to
the R&D organization, as it happens in the case of GM, others to the
strategy organization, as is the case with Ford, and yet others to a specific business unit as was the case until recently at BMW. Good coordination continues with clarity of mission for the Innovation Outpost.
For example, as we saw in Chapter 7, some of the Silicon Valley-based
automotive Innovation Outposts are actually acting as scouts for the
corporate parent's R&D organization, others conduct actual technology
development, indicating that they have a technology-only mission, and
yet others are part of a business development function.

If startups are to play an important role in the way the automotive
incumbents innovate and succeed with big data and machine intelligence, the Innovation Outpost must ensure that there is constant value
exchange between the startups and the corporation. In this respect, the
Innovation Outpost is the critical link. Three basic stakeholders (shown
in Figure 17) need to come together for this to be accomplished:

1. The corporate leadership (CEO, senior executive team, board of directors).
2. The Innovation Outpost.
3. The startup ecosystem in each innovation cluster, e.g., startups, entrepreneurs, investors, etc.

Take one of these stakeholders away, and the entire effort weakens substantially and may even collapse. The automotive incumbent must not see value only in investing in a startup or enabling it to use its brand by announcing a partnership. Today, most startups trying to work with the automotive industry complain that there is only a one-way flow of information from them to the incumbents that express an interest in working with them. In other words, there is no value exchange.

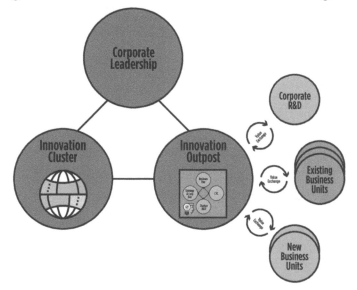

FIGURE 17: THE INNOVATION OUTPOST IS THE CRITICAL VALUE EXCHANGE LINK BETWEEN THE CORPORATION AND THE STARTUP ECOSYSTEM. FIGURE COURTESY OF EVANGELOS SIMOUDIS

Particularly as part of the data-sharing culture, in addition to the value exchange between the automotive incumbent and the startups, it is also important that the incumbent's Innovation Outposts collaborate with Innovation Outposts of other members of the automotive value chain, as shown in Figure 18.

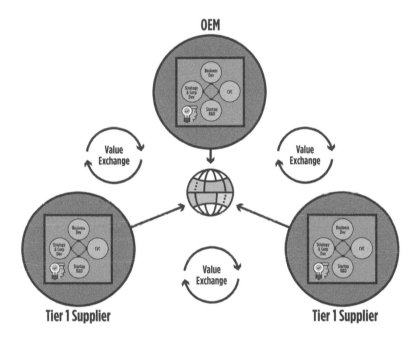

FIGURE 18: VALUE EXCHANGE AMONG INNOVATION OUTPOSTS OF OEMS AND THEIR TIER 1 SUPPLIERS. FIGURE COURTESY OF EVANGELOS SIMOUDIS

Recommendation 4: Establish and execute a long-term strategy to attract the right big data and machine intelligence talent and create the right culture.

The enabling technologies for ACE vehicles and on-demand Mobility Services, e.g., software, mobile applications, big data, etc., are in high-demand by a variety of industries. In order to attract and ultimately hire employees with these areas of expertise, automotive incumbents will need to establish and execute on a differentiated and long-term talent-acquisition strategy. As part of this strategy, they will need to form separate organizations, outside their core operations, with a culture that supports agility, risk-taking, and entrepreneurial thinking at every level. Moreover, by establishing separate organizations as Ford, BMW, Daimler, and GM have already done they will be able to retain the parts of their existing corporate culture that defines each of their companies.

As part of their big data talent-acquisition strategy, they will need to determine how many employees they will be able to hire in the open market and how much of the talent gap they will close through a) acquisitions (including opportunities to first invest in a startup and later acquire it), b) retraining employees who already have a quantitative background, c) providing incubation opportunities to the most entrepreneurial of these employees to experiment with potentially disruptive big data solutions and d) establishing relations with universities that run strong data science and machine intelligence programs. An interesting approach in this area is taken by Toyota. In addition to the company's Silicon Valley-based Information Technology Center that is working on big data, the company is in discussions to buy Google's Boston Dynamics division that has a strong big data team, and the company has committed $1B to create a center for artificial intelligence, and robotics research that will also focus on big data.

As they implement their talent-acquisition strategy, automotive industry incumbents must appreciate that they are competing for candidate employees that are also sought by corporations in many other industries, as well as a large number of well-financed startups. These candidates want to work for companies that have a critical mass of employees in their area of focus, e.g., big data, and be located in the innovation clusters where the automotive industry has not had significant presence. For example, every year US universities award only about 2000 PhDs in computer science. Companies like Verizon,[166] Ericsson,[167] and Samsung[168] have already realized that in this war for big data talent, they must fight with startups, Tesla, Google, Facebook, Amazon, Uber, Baidu, and several others. For this reason, these companies are making significant investments in innovation clusters such as the Bay Area, Cambridge, Austin, Herzliya, Shanghai, and Berlin in order to have early access to this talent. In these areas they are opening fully functioning Innovation Outposts and frequently establish large centers in order to create the appropriate critical mass in their areas of strategic interest, such as big data. The automotive industry has to follow the same playbook. However, automotive incumbents will also need to provide the well-articulated reasons of why a candidate should join them instead of one of the other companies looking for exactly the same set of big data skills. Only by creating the critical mass with the new thinkers of big data will they be in the position to disrupt and impact the thinking and decisions of the business units that are typically located away from the innovation clusters.

Recommendation 5: Use scale as an advantage to enhance the strategic big data efforts with those of the startup ecosystem.

While it is extremely attractive to think big and try to disrupt an entire industry, startups realize that it is hard to break into and succeed in the automotive industry. The industry requires high investment and ability to scale while its risk tolerance remains low. Entering the Mobility Services area may have lower capital requirements but, as a result, it has low barriers to entry, resulting in many competitors that ultimately have difficulty differentiating their solutions and thus require large investments in order to scale globally with no guarantees for success. For example, Uber recently raised an additional $3.5B to support its expansion plans and its plan to create its own fleet of self-driving cars. Underlining the risks of having low barriers to entry, Uber has recently been forced to sell its China operations to Didi Chuxing[169] after determining, despite spending over $2B, it cannot win in the Chinese market.

Incumbents must use their scale and global reach to their advantage. For example, many of the millions of cars they have already sold and will continue to sell on an annual basis can become data collection platforms. This is a strategy that is already being exploited by Tesla. Incumbent automakers have an advantage, should they choose to exploit it, because of the orders of magnitude more vehicles they have on the road compared to Tesla and other newcomers. To capitalize on this advantage, they must determine how to utilize the sensors already installed in their vehicles, identify the additional sensors they will need to incorporate in future models, establish data privacy and data-sharing policies, as well as implement an architecture such as the one described in Chapter 8. Innovation-minded automotive industry incumbents that embrace big data as a strategic imperative must then be in an excellent position to convince startups with important Intellectual Property (IP) and expertise that by collaborating effectively around their respective data assets they will be able to both succeed. This type of collaboration will enable a startup to remain focused on its core competence and maintain its capital efficiency while leveraging the incumbent's industry knowledge, well-established processes, networks, and products. Two examples for what may be possible for big data startups as they consider partnerships with automakers come from NVIDIA and MobileEye. NVIDIA started focusing on systems for autonomous vehicles and over several years established partnerships with several automakers that are now working on such vehicles. Its products are now part of its partners' supply chains. Similarly MobileEye developed critical systems for car safety and focused on autonomous cars,[170] its most well-known customer to date being Tesla. Collaborative partnerships can take several forms. They may start with a venture investment, the opportunity to jointly develop new IP, or an agreement to make the startup part of the automakers, big data ecosystem, not unlike what IBM is currently doing with its Watson cognitive platform or GE with its Predix platform.

10.0
CONCLUSION

We live in extraordinary times. Concepts and ideas we used to only dream about or watch in science fiction films just a few short years ago, such as driverless cars, we are now able to start experiencing in our daily lives. And the speed with which such concepts are becoming reality continues to increase. While as consumers we are in awe at the pace of innovation, many industries are already being disrupted or are at high risk of being disrupted in the very near future. Disruptive innovations such as Autonomous Connected Electrified vehicles coupled with novel on-demand Mobility Services are leading to a hybrid model that combines car ownership with car access are creating such a risk for the incumbent automotive industry. This risk is starting to become evident in urban areas, and specifically megacities, where large segments of the world's population are moving and where such innovations are being tested to address important problems. Even if these problems are ultimately not fully solved through the innovations in personal mobility, the risk of being disrupted by or the opportunity to capitalize on these innovations is one that the incumbent automotive industry cannot ignore.

In order to respond to the emerging personal mobility innovations, automotive industry incumbents have been taking several steps. They are increasing their R&D investments, making acquisitions, and starting to work with the ecosystems of innovation clusters such as Silicon Valley, Israel, and China.

However, automotive incumbents need to adopt a long-term strategic approach to address less reactively the driverless future. The book identifies big data as a key area for incumbents to focus. It offers recommendations on how to succeed in this area by adopting a strategy that blends corporate innovation efforts with innovations created by startups, and a play book developed through the experiences of startup ecosystems. Adopting these recommendations will require incumbents to make tough financial, organizational, and technology decisions and take actions faster. Many of these decisions will require painful cultural changes.

The automotive industry as we know it today has emerged from a group of startups, many of which were established in Detroit, a city that at the turn of the 20th century was an innovation cluster, quite similar to today's Silicon Valley. Over the years, the industry has weathered many challenges, several due to economic downturns, most recently in 2009, others due to regulation, and globalization. It always found a way to succeed and come out stronger. However, its overall business model has never been challenged for the past 100-plus years of its existence in the way it is being challenged today by both startups and large corporations, that are newcomers to the automotive industry. If it is to avoid being disrupted by and continue to disrupt in the coming driverless and on-demand mobility future, it will need to make fundamental changes and among a few other strategic options recognize the profound and long-term opportunities provided by the broad exploitation of big data.

BIBLIOGRAPHY

Chapter 1

Chapter 2

1. http://www.businessinsider.com/tesla-raises-nearly-15-billion-in-fresh-capital-2016-5

2. https://www.engadget.com/2016/01/04/faraday-future-zero-1/

3. http://www.businessinsider.com/car-companies-of-the-world-2015-2

4. http://marketrealist.com/2015/02/intense-competition-leads-low-profit-margins-automakers/

5. http://www.Linkedin.com/in/loukerner

6. https://www.nada.org/nadadata/

7. http://www.autonews.com/article/20130926/OEM06/130929919/ford-buys-small-software-startup-to-bolster-connected-car-offerings

Chapter 3

8. http://esa.un.org/unpd/wup/highlights/wup2014-highlights.pdf

9. http://www.marketplace.org/2016/08/02/world/more-companies-trade-suburbia-city

10. http://cusp.nyu.edu/

11. http://www.tomtom.com/en_gb/trafficindex/

12. http://www.cnn.com/2015/01/27/asia/asia-air-pollution-haze/index.html

13. http://www.air-quality.org.uk/11.php

14. https://www.bloomberg.com/graphics/2015-whats-warming-the-world/?1435237748266=1

15. https://www.bloomberg.com/view/articles/2015-04-20/in-new-millennium-no-jobs-for-millennials

16. http://www.economist.com/news/leaders/21573104-internet-everything-hire-rise-sharing-economy

17. http://www.web-strategist.com/blog/research/

18. http://www.tandfonline.com/doi/abs/10.1080/15389588.2012.696755

19. http://www.businessinsider.com/everything-we-thought-about-millennials-not-buying-cars-was-wrong-2016-3

20. https://www.nap.edu/read/12794/chapter/1

21. http://repository.cmu.edu/cgi/viewcontent.cgi?article=2874&context=compsci

22. https://en.wikipedia.org/wiki/DARPA_Grand_Challenge

23. https://en.wikipedia.org/wiki/Sebastian_Thrun

24. https://www.google.com/selfdrivingcar/

25. http://www.nytimes.com/2015-04-03/automobiles/ semiautonomous-driving-arrives-feature-by-feature.html?_r=0

26. http://www.driverless-future.com/?page_id=384

27. https://www.cbinsights.com/blog/autonomous-driverless-vehicles-corporations-list/

28. http://bit.ly/rb_AV_Q32016

29. https://en.wikipedia.org/wiki/Cloud_robotics

30. https://www.engadget.com/2016/01/04/faraday-future-zero-1/

31. https://www.wired.com/2016/06/teslas-plan-rule-auto-industry-app-purchases/

32. http://www.techradar.com/news/world-of-tech/ connected-cars-a-cyber-security-nightmare-on-wheels-1277541

33. http://www.wsj.com/articles/tech-companies-still-trusted-more-on-autonomous-car-development-1469795264

34. http://www.gearthblog.com/blog/ archives/2014/04/google-earth-imagery.html

35. http://360.here.com/2016/09/26/here-first-to-unveil-services-from-sensor-data-of-multiple-car-brands/

36. https://techcrunch.com/2016/04/13/nauto-raises-12-million-for-driverless-car-technology-thats-street-legal-today/

37. http://www.theverge.com/2016/6/6/11866868/ comma-ai-george-hotz-interview-self-driving-cars

38. http://www.theverge.com/2016/7/31/12338268/ uber-maps-investment-500-million

39. http://time.com/3851639/uber-here-nokia-maps/

40. http://www.theverge.com/2015/3/3/8145035/ ubers-first-confirmed-acquisition-is-a-mapping-company

41. https://www.wired.com/2014/12/nokia-here-autonomous-car-maps/

42. http://www.theverge.com/2016/4/1/11346710/ amazon-microsoft-here-autonomous-car-tech-investment

43. http://instapage.cbinsights.com/research-google-acquisitions

44. http://money.cnn.com/2015/09/28/technology/ apple-maps-hopstop-mapsense/index.html

45. http://mashable.com/2015/10/14/tesla-high-precision-digital-maps/

46. http://www.economist.com/news/science-and-technology/21696925-building-highly-detailed-maps-robotic-vehicles-autonomous-cars-reality

47. http://www.theatlantic.com/technology/ archive/2012/09/how-google-builds-its-maps-and-what-it-means-for-the-future-of-everything/261913/

48. http://www.usatoday.com/story/tech/ news/2016/05/24/google-maps-ads/84854240/

49. https://www.fastcompany.com/3050250/what-makes-uber-run

50. http://www.latimes.com/business/technology/la-fi-0105-lyft-growth-20160105-story.html

51. https://en.wikipedia.org/wiki/Didi_Chuxing

52. http://www.cleanfleetreport.com/best-car-sharing/

53. http://www.vtpi.org/multimodal_planning.pdf

54. http://www.pwc.com/gx/en/industries/ transportation-logistics/tl2030.html

55. https://www2.deloitte.com/tr/en/pages/public-sector/articles/digital-age-transportation-article.html

56. http://techcrunch.com/2016/05/03/moovit-transit-app-integrates-with-uber/

57. http://www.mckinsey.com/industries/high-tech/our-insights/disruptive-trends-that-will-transform-the-auto-industry

58. http://www.curbed.com/2016/8/8/12404658/autonomous-car-future-parking-lot-driverless-urban-planning

59. http://www.rmi.org/peak_car_ownership

60. https://www.quora.com/How-many-commuter-buses-do-Google-Facebook-Apple-and-other-Silicon-Valley-buses-run-for-their-employees-every-day

61. http://www.vta.org/sfc/servlet.shepherd/version/download/068A0000001FZYQ

62. http://www.slate.com/articles/technology/future_tense/2016/06/the_self_driving_car_generation_gap.html

Chapter 4

63. https://en.wikipedia.org/wiki/Big_data

64. http://techcrumch.com/2015/11/26/machine-intellegence-in-the-real-world/

65. https://www.oreilly.com/ideas/from-insight-as-a-service-to-insightful-applications

66. https://www.wired.com/2016/03/self-driving-cars-wont-work-change-roads-attitudes/

67. http://www.computerworld.com/article/2484219/emerging-technology/self-driving-cars-could-create-1gb-of-data-a-second.html

68. http://www.reuters.com/article/us-tesla-investigation-idUSKCN0ZG2ZC

69. http://www.forbes.com/sites/samabuelsamid/2016/08/16/2017-audis-will-talk-to-some-traffic-signals-kicking-off-vehicle-to-infrastructure-communications

70. http://www.futurecities.ethz.ch/project/engaging-active-mobility/

71. http://www-formal.stanford.edu/jmc/whatisai/

72. http://www.rollingstone.com/culture/features/inside-the-artificial-intelligence-revolution-a-special-report-pt-1-20160229

73. http://machinelearningmastery.com/a-tour-of-machine-learning-algorithms/

74. http://www.nature.com/nature/journal/v518/n7540/full/nature14236.html

75. https://en.wikipedia.org/wiki/Artificial_neural_network

76. http://electrek.co/2016/04/11/google-self-driving-car-tesla-autopilot/

77. http://www.newsweek.com/ubers-plan-map-world-and-fill-it-self-driving-cars-google-maps-485885

78. http://www.theverge.com/2016/2/1/10892020/google-self-driving-simulator-3-million-miles

79. https://www.nytimes.com/interactive/2016/06/06/automobiles/autonomous-cars-problems.html?_r=0

80. https://www.nytimes.com/2016/06/24/technology/should-your-driverless-car-hit-a-pedestrian-to-save-your-life.html?_r=0

81. http://www.driverless-future.com/?page_id=384

Chapter 5

82. http://www.forbes.com/sites/davidnicholson/2014/11/09/inside-tesla-a-rare-glimpse-of-electric-carmakers-culture/

83. https://www.quora.com/What-are-some-of-Tesla-Motors-major-innovations

84. http://www.tradegecko.com/blog/tesla-custom-built-supply-chain

85. http://mashable.com/2015/03/15/tesla-elon-musk-battery-range/#RRbmik2Hbgq9

86. http://files.shareholder.com/downloads/abea-rcw8x01/521457336x8005399x858516/f5099-faf-ba73-4263-8e16-de1fac0babdf/q3_15_sharholder_letter.pdf

87. http://www.reuters.com/article/us-tesla-masterplan-idUSKCN1002Q4

88. http://www.wsj.com/articles/tesla-ceo-sees-fully-autonomous-car-ready-in-five-or-six-years-1410990887

89. http://fortune.com/2016/11/20/tesla-urban-navigation-video/

90. http://electrek.com/2016/04/21/elon-musk-tesla-mobility-service

91. http://www.businessinsider.com/tesla-driverless-ridesharing-plans-could-take-on-uber-2016-10

92. http://online.wsj.com/news/articles/SB1000142412788732437400457821712143332386

93. http://www.theverge.com/2016/6/28/12051220/zipcar-pay-per-mile-car-sharing

94. http://www.forbes.com/sites/aswathdamodaran/2014/06/10/a-disruptive-cab-ride-to-riches-the-uber-payoff/#6e86a2aa64b8

95. https://www.oreilly.com/ideas/ubers-case-for-incremental-processing-on-hadoop

96. https://recode.net/2015/02/02/uber-tries-to-speed-up-arrival-of-robot-cars-tells-drivers-not-to-worry-qa/

97. https://www.bloomberg.com/news/features/2016-08-18/uber-s-first-self-driving-fleet-arrives-in-pittsburgh-this-month-is06r7on

98. https://www.fastcompany.com/3050250/what-makes-uber-run

99. https://dupress.deloitte.com/dup-us-en/focus/future-of-mobility/transportation-technology.html

100. https://techcrunch.com/2015/02/02/uber-opening-robotics-research-facility-in-pittsburgh-to-build-self-driving-cars/

101. http://bits.blogs.nytimes.com/2015/06/29/uber-to-acquire-mapping-technology-and-know-how-from-microsoft/

102. http://www.theatlantic.com/technology/archive/2016/07/ubers-latest-push-toward-driverless-cars/493271/

103. http://www.nytimes.com/2016/08/19/technology/uber-self-driving-cars-otto-trucks.html

104. http://spectrum.ieee.org/transportation/advanced-cars/meet-zoox-the-robotaxi-startup-taking-on-google-and-uber

105. http://www.techtimes.com/articles/66684/20150707/google-sends-self-driving-lexus-suv-to-test-the-roads-in-austin.htm

106. http://www.autoblog.com/2016/05/20/google-fca-minivans-report/

107. http://www.usatoday.com/story/tech/2015/06/25/googles-autonomous-pod-cars-hit-the-road/29292169/

108. http://www.cbc.ca/news/business/uber-five-things-1.3351307

109. http://www.salon.com/2016/01/06/uber_fail_why_the_start_up_giant_stumbled_in_europe_and_how_it_could_happen_in_the_u_s/

110. http://www.businessinsider.com/why-uber-failed-in-china-2016-8

111. http://www.telegraph.co.uk/technology/google/12095898/Googles-driverless-cars-needed-hundreds-of-human-interventions-to-prevent-accidents-and-failures.html

112. https://mitsloan.mit.edu/learingedge/casedocs/14-153.robin chase and zipcar.final.pdf

113. https://www.gilderlehrman.org/history-by-era/politics-reform/essays/motor-city-story-detroit

Chapter 6

114. http://www.businessinsider.com/how-google-bought-waze-the-inside-story-2015-8

115. http://wp.me/p4vHBq-z

116. http://www.reuters.com/article/us-bmw-electric-idUSKCN0YO1YZ

117. http://www.mckinsey.com/business-functions/strategy-and-corporate-finance/our-insights/enduring-ideas-the-three-horizons-of-growth

118. https://foundry.unilever.com/welcome

119. https://techcrunch.com/2015/09/29/the-online-buy-now-pay-later-service-klarna-adds-amex-as-a-partner/

120. https://www.brookings.edu/research/the-new-cluster-moment-how-regional-innovation-clusters-can-foster-the-next-economy

121. http://wp.me/p4vHBq-6b

122. https://www.qualcomm.com/invention/cognitive-technologies/robotics

123. https://twitter.com/kmelrobotics

124. https://qualcommventures.com/portfolio/3d-robotics

125. http://qualcomm.com/news/onq/2015/07/16/qualcomm-robotics-accelerator-backs-10-robotic-start-ups

126. http://www.techstars.com/

127. http://features.marketplace.org/priceofprofits/

128. http://reports.weforum.org/future-of-jobs-2016/

129. https://www.engadget.com/2014/07/17/tesla-motors-us-sales/

130. http://www.autoalliance.org/auto-jobs-and-economics/2015-jobs-report

131. https://www.nada.org/nadadata/

Chapter 7

132. http://www.strategyand.pwc.com/reports/2015-global-innovation-1000-media-report

133. http://www.strategyand.pwc.com/reports/2015-global-innovation-1000-media-report

134. http://www.strategyand.pwc.com/reports/2015-global-innovation-1000-media-report

135. http://www.strategyand.pwc.com/media/file/the-2014-global-innovation-1000_media-report.pdf

136. https://www.bcgperspectives.com/content/articles/growth-lean-manufacturing-innovation-in-2015/

137. https://www.bcgperspectives.com/content/
      articles/innovation_growth_digital_economy_
      innovation_in_2014/

Chapter 8

138. http://www.nytimes.com/2010/02/05/
      technology/05electronics.html

139. http://www.makeuseof.com/tag/obd-ii-port-used/

140. https://news.samsung.com/global/samsung-
      ushers-in-a-new-era-of-driving-experience-with-
      samsung-connect-auto

141. http://dataconomy.com/2015/08/how-big-data-
      brought-ford-back-from-the-brink/

142. http://www.zdnet.com/article/ford-ceo-
      fields-on-autonomous-cars-big-data-
      tesla/#ftag=RSSbaffb68

143. http://www.bmwblog.com/2016/06/27/bmw-audi-
      mercedes-purchase-maps/

144. http://fortune.com/2016/07/15/ford-motor-3d-
      mapping/

145. http://gmauthority.com/blog/2015/01/35-percent-
      of-onstar-remotelink-requests-are-for-remote-
      start-plus-other-interesting-stats-about-the-
      onstar-app/

146. https://en.wikipedia.org/wiki/OnStar

147. http://linkis.com/cleantechnica.com/BMW_i_
      Ventures_Inves_1.html

148. http://bigdata-madesimple.com/ibm-analytics-
      improve-bmw-auto-quality/

149. http://economictimes.indiatimes.com/automobiles/
      mercedes-benz-bets-on-50-data-scientists-at-its-
      bangalore-unit-to-aid-lower-component-failures/
      articleshow/40074573.cms

150. http://blogs.wsj.com/cio/2013/05/15/how-gm-lost-
      a-mainframe-but-gained-an-it-department/

151. http://fortune.com/2013/10/16/how-tesla-
      autopilot-learns/?xid-entrepreneur

152. https://blogs.nvidia.com/blog/2016/01/05/eyes-
      on-the-road-how-autonomous-cars-understand-
      what-theyre-seeing/#sthash.tPfXhkKT.dpuf

153. http://bgr.com/2015/10/20/tesla-software-updates-
      the-future-of-cars/

154. http://sandhill.com/article/huge-cybersecurity-
      market-protecting-cars-from-being-hacked/

155. https://www.munichre.com/site/mram/get/
      documents_E732654315/mram/assetpool.
      mr_america/PDFs/3_Publications/RAND_
      Autonomous%20Vehicle%20Guide%20for%20
      Policymakers.pdf

156. https://securityledger.com/2016/07/
      auto-industry-publishes-best-practices-for-
      cybersecurity/?ReillyBrennanFoT

157. http://fortune.com/2016/05/13/uber-apple-china-
      didi/

158. https://www.wired.com/2014/06/the-nissan-van-
      taxi-that-will-rule-nycs-streets-is-actually-great/

159. http://www.reuters.com/article/us-gm-lyft-
      investment-idUSKBN0UI1A820160105

160. https://www.bloomberg.com/news/
      articles/2016-05-24/toyota-to-invest-in-uber-and-
      team-up-on-auto-leasing-program

161. https://techcrunch.com/2016/05/24/vw-invests-
      300m-in-uber-rival-gett-in-new-ride-sharing-
      partnership

162. http://www.reuters.com/article/us-bmw-
      investment-idUSKCN0YG247

163. https://www.sbdautomotive.com/files/sbd/
      pdfs/55lib.pdf?dm_i=1icz,259in,89xtkx,7r2gy,1

164. http://www.mckinsey.com/industries/automotive-
      and-assembly/our-insights/monetizing-car-data

Chapter 9

165. http://www.sustainablebrands.com/news_
      and_views/business_models/hannah_furlong/
      bmw_daimler_expand_car-sharing_services_us

166. http://www.bizjournals.com/sanjose/
      news/2014/06/03/verizon-reaches-for-
      highersilicon-valley-profile.html?page=all

167. http://www.fiercewireless.com/tech/story/ericsson-
      boosting-presence-investment-silicon-valley-new-
      rd-campus/2014-05-28

168. http://venturebeat.com/2015/01/14/samsung-on-
      track-to-finish-its-futuristic-silicon-valley-campus-
      before-apple-google-or-facebook/

169. http://www.wsj.com/articles/china-s-didi-
      chuxing-to-acquire-rival-uber-s-chinese-
      operations-1470024403

170. http://www.nasdaq.com/article/car-safety-ipo-
      mobileye-ties-future-to-tesla-others-cm472726

Chapter 10

INDEX

ABOUT THE AUTHOR

Evangelos Simoudis is a recognized expert on big data strategies and corporate innovation. He has been working in Silicon Valley for 25 years as a venture investor, entrepreneur, and corporate executive. He is the co-founder and Managing Director of Synapse Partners, a venture firm that invests in early stage startups developing big data applications, and an advisor to global corporations on big data strategies and Startup-Driven Innovation.

Evangelos has also served as Managing Director at Trident Capital, and Partner at Apax Partners. In 2012 and 2014 he was named top investor in online advertising. Prior to his venture and advisory career Evangelos served as President and CEO of Customer Analytics, and as Vice President of Business Intelligence at IBM.

Evangelos serves on advisory boards for the Center for Information Science and Technology at Caltech, the International School of Business at Brandeis University, the Center for Urban Science and Progress at New York University, and the Autonomous Vehicle Task Force of SAFE.

Evangelos earned a PhD in computer science from Brandeis University, and a BS in electrical engineering from Caltech.

CPSIA information can be obtained
at www.ICGtesting.com
Printed in the USA
LVHW050919220519
618545LV00007B/21/P